PHYSICS

IB DIPLOMA PROGRAMME

David Homer

OXFORD
UNIVERSITY PRESS

OXFORD
UNIVERSITY PRESS

Great Clarendon Street, Oxford, OX2 6DP, United Kingdom

Oxford University Press is a department of the University of Oxford. It furthers the University's objective of excellence in research, scholarship, and education by publishing worldwide. Oxford is a registered trade mark of Oxford University Press in the UK and in certain other countries

© Oxford University Press 2019

Acknowledgements

Photo credits:

Cover image: Alamy Stock Photo; **p8**: Hxdbzxy/Shutterstock; **p2**: Mmaxer/Shutterstock; **p16**: Germanskydiver/Shutterstock; **p32**: Schankz/Shutterstock; **p53**: Giphotostock/Science Photo Library; **p42**: Zffoto/Shutterstock; **p60**: Ase/Shutterstock; **p74**: SIHASAKPRACHUM/ Shutterstock; **p81**: ALASTAIR PHILIP WIPER /Science Photo Library; **p92**: Saman527/Shutterstock; **p100**: Ella Hanochi/Shutterstock; **p114**: Awe Inspiring Images/Shutterstock; **p124**: Electrical Engineer/Shutterstock; **p136**: PHILIPPE PLAILLY/Science Photo Library; **p148**: Raimundas/ Shutterstock; **p154**: PatinyaS/Shutterstock; **p172**: Brize99/Shutterstock; **p190**: PeterPhoto123/Shutterstock; **p210**: Vadim Sadovski/Shutterstock; **p230**: CandMe/Shutterstock;

All other photos by David Homer.

Artwork by Aptara Corp. and OUP. Index by James Helling.

The publisher would like to thank the International Baccalaureate for their kind permission to adapt questions from past examinations. The questions adapted for this book, and the corresponding past paper references, are summarized here:

Sample student answer, **p13**: N16 TZ0 HL P2 Q1**(a)**; Example 2.2.1, **p14**: N09 SL P2 Q2**(a)(iii)**; Example 2.2.4, **p16**: M09 TZ1 SL P2 Q2**(b),(c)**; Sample student answer, **p16**: M17 TZ2 HL P2 Q1**(d)**; Example 2.3.1, **p18**: Adapted from M10 TZ2 SL P2 Q2**(a),(b)(i)**; Sample student answer, **p20**: M17 TZ2 HL P2 Q1**(a)**; Example 2.4.2, **p22**: Adapted from N10 SL P2 Q2**(b) (i),(c)**; Example 2.4.4, **p23**: M09 TZ1 HL P2 Q2**(b)(i),(ii),(c)**; Example 3.1.3, **p27**: Adapted from M09 HL Q2**(a)**; Sample student answer, **p28**: M16 HL Q3**(a),(b)**; Sample student answer, **p32**: N16 HL Q3**(b)**; Sample student answer, **p39**: Adapted from M16 HL Q4**(a),(b)**; Sample student answer, **p42**: N16 HL Q5**(b),(c)**; Sample student answer, **p46**: M17 TZ2 HL Q4**(a),(b) (i)**; Sample student answer, **p51**: M17 TZ2 HL Q4**(d)**; Sample student answer, **p54**: M17 TZ2 HL Q6**(a)**; Sample student answer, **p58**: N16 HL Q8**(a)** and M16 SL Q5**(c)**; Sample student answer, **p61**: M16 SL Q5**(b)**; Sample student answer, **p64**: M16 SL Q5**(d)**; Sample student answer, **p68**: M17 TZ2 SL Q1**(d)**; Sample student answer, **p70**: M16 HL Q2 and M16 SL Q2; Example 7.1.1, **p72**: M11 TZ2 P3 SL B2; Sample student answer, **p75**: M16 SL P2 Q6**(b)**; Sample student answer, **p78**: M16 SL P2 Q6**(a)**; Sample student answer, **p82**: N16 HL P2 Q4; Example 8.1.1, **p85**: Adapted from M10 TZ1 HL P1 Q37; Sample student answer, **p87**: M17 TZ2 SL P2 Q2**(b),(c)**; Sample student answer, **p90**: N16 HL P2 Q10**(a)**; Practice problem 4, **p91**: M11 TZ2 HL P2 Q2**(d)**; Sample student answer, **p95**: M18 TZ2 HL P2 Q1**(d)**; Sample student answer, **p100**: M17 TZ2 HL P2 Q4**(a),(b)**;Example 9.5.1, **p103**: M14 TZ2 SL P3 Q2**(b)**; Sample student answer, **p104**: N16 HL P2 Q6**(a)**; Sample student answer, **p108**: N16 HL P2 Q7**(a)**; Sample student answer, **p113**: M17 TZ2 HL P2 Q8**(a),(b)(i)**; Sample student answer, **p117**: M14 TZ2 HL P2 Q5**(b)**; Example 11.2.2, **p119**: Adapted from M10 TZ1 HL P1 Q26; Sample student answer, **p122**: N16 HL P2 Q10**(c)**; Sample student answer, **p126**: M16 HL P2 Q7; Sample student answer, **p132**: N16 HL P2 Q11**(a)**; Sample student answer, **p138**: M18 TZ2 HL P2 Q9**(d)**; Sample student answer, **p143**: M18 TZ2 SL P3 Q2; Sample student answer, **p144**: M16 SL P3 Q2; Practice problem 1, **p145**: Adapted from M06 TZ2 HL P2 Q1; Sample student answer, **p148**: M16 SL P3 Q4; Example A.2.1, **p150**: M11 TZ1 SL P3 Q2**(a)**; Sample student answer, **p151**: M16 SL P3 Q5**(a)**; Sample student answer, **p156**: M16 SL P3 Q6**(b)**; Sample student answer, **p158**: M17 TZ2 HL P3 Q6; Sample student answer, **p162**: M18 TZ2 HL P3 Q7; Sample student answer, **p167**: Adapted from M17 TZ2 SL P3 Q8; Sample student answer, **p172**: M17 TZ2 HL Q9**(a),(b)**; Sample student answer, **p177**: M17 TZ2 HL Q10**(a),(b)**; Sample student answer, **p180**: M17 TZ2 HL Q11; Sample student answer, **p187**: N16 HL P3 Q16**(b),(c),(d)**; Example C.2.2, **p190**: M12 TZ2 P3 G2; Sample student answer, **p192**: N16 HL P3 Q17; Sample student answer, **p195**: N16 HL P3 Q18; Example C.4.2, **p199**: Adapted from M12 HL P3 I2**(b)**; Sample student answer, **p200**: M18 TZ2 HL P3 Q15**(a),(b)**;Sample student answer, **p204**: N16 HL P3 Q21**(a),(b)**; Sample student answer, **p209**: N16 HL P3 Q22; Sample student answer, **p213**: M16 HL P3 Q19; Sample student answer, **p216**: N16 HL P3 Q24**(a),(b)**; Sample student answer, **p219**: N16 HL P3 Q25; Practice paper 1, Q17, **p228**: Adapted from M10 TZ1 P1 Q20; Practice paper 1, Q28, **p229**: M10 TZ1 P1 Q39; Practice paper 1, Q39, **p230**: M15 TZ2 P1 Q29; Practice paper 2, Q3, **p231**: M11 TZ1 P2 B3**(c),(d)**; Practice paper 2, Q7, **p232**: M12 TZ2 P3 A5**(b)**; Practice paper 2, Q8, **p233**: M12 TZ2 P2 A6**(b)**; Practice paper 3, Q3, **p235**: N16 HL P3 Q4; Practice paper 3, Q4, **p235**: M17 TZ2 P3 Q5; Practice paper 3, Q14, **p237**: M17 TZ2 P3 Q12**(a),(c)**; Practice paper 3, Q16, **p237**: M10 TZ2 P3 F4**(a),(c)**; Practice paper 3, Q17, **p238**: M10 TZ2 P3 I3; Practice paper 3, Q20, **p238**: M12 TZ1 P3 E1**(a),(e)**; Practice paper 3, Q21, **p238**: M12 TZ2 P3 E1**(b),(c)**; Practice paper 3, Q22, **p239**: M10 TZ2 P3 E2; Practice paper 3, Q24, **p239**: M09 TZ1 P3 E3.

Contents

Answers to questions and exam papers in this book can be found on your free support website. Access the support website here:

www.oxfordsecondary.com/ib-prepared-support

INTRODUCTION

This book provides full coverage of the IB diploma syllabus in Physics and offers support to students preparing for their examinations. The book will help you revise the study material, learn the essential terms and concepts, strengthen your problem-solving skills and improve your approach to IB examinations. The book is packed with worked examples and exam tips that demonstrate best practices and warn against common errors. All topics are illustrated by annotated student answers to questions from past examinations, which explain why marks may be scored or missed.

A separate section is dedicated to data-based and practical questions, which are the most distinctive feature of the syllabus. Numerous examples show how to tackle unfamiliar situations, interpret and analyse experimental data, and suggest improvements to experimental procedures. Practice problems and a complete set of IB-style examination papers provide further opportunities to check your knowledge and skills, boost your confidence and monitor the progress of your studies. Full solutions to all problems and examination papers are given online at **www.oxfordsecondary.com/ib-prepared-support.**

As any study guide, this book is not intended to replace your course materials, such as textbooks, laboratory manuals, past papers and markschemes, the IB Physics syllabus and your own notes. To succeed in the examination, you will need to use a broad range of resources, many of which are available online. The author hopes that this book will navigate you through this critical part of your studies, making your preparation for the exam less stressful and more efficient.

DP Physics assessment

All standard level (SL) and higher level (HL) students must complete the internal assessment and take three papers as part of their external assessment. Papers 1 and 2 are usually sat on one day and Paper 3 a day or two later. The internal and external assessment marks are combined as shown in the table at the top of page V to give your overall DP Physics grade, from 1 (lowest) to 7 (highest).

Overview of the book structure

The book is divided into several sections that cover the internal assessment, core SL and additional higher level (AHL) topics, data-based and practical questions, the four options (A–D) and a complete set of practice examination papers.

The largest section of the book, **core topics**, follows the structure of the IB diploma physics syllabus and covers all *understandings* and *applications and skills* assessment statements. Topics 1–8 contain common material for SL and HL students, while topics 9–12 are intended for HL students only. The *nature of science* concepts are also mentioned where applicable.

The **data-based and practical questions** section (Chapter 13) provides a detailed analysis of problems and laboratory experiments that often appear in section A of paper 3. Similar to core topics, the discussion is illustrated by worked examples and sample scripts, followed by IB-style practice problems.

The **options** section reviews the material assessed in the second part of paper 3. Each of the four options is presented as a series of SL and AHL subtopics.

The **internal assessment** section outlines the nature of the investigation that you will have to carry out and explains how to select a suitable topic, collect and process experimental data, draw conclusions and present your report in a suitable format to satisfy the marking criteria and achieve the highest grade.

The final section contains IB-style **practice examination papers 1, 2 and 3**, written exclusively for this book. These papers will give you an opportunity to test yourself before the actual exam and at the same time provide additional practice problems for every topic of core and options material.

The answers and solutions to all practice problems and examination papers are given online at **www.oxfordsecondary.com/ib-prepared-support.** Blank answer sheets for examination papers are also available at the same address.

Assessment overview

Assessment	Description	Topics	SL marks	SL weight	HL marks	HL weight
Internal	Experimental work with a written report	—	24	20%	24	20%
Paper 1	Multiple-choice questions	Core: 1–8 (SL) 1–12 (AHL)	30	20%	40	20%
Paper 2	Short- and extended-response questions		50	40%	90	36%
Paper 3	Section A: data-based and practical questions	—	15	20%	15	24%
	Section B: short- and extended-response questions	Option of your choice	20		30	

The final IB diploma score is calculated by combining grades for six subjects with up to three additional points from *theory of knowledge* and *extended essay* components.

Command terms

Command terms are pre-defined words and phrases used in all IB Physics questions and problems. Each command term specifies the type and depth of the response expected from you in a particular question. For example, the command terms *state, outline, explain* and *discuss* require answers with increasingly higher levels of detail, from a single word, short sentence or numerical value ("state") to comprehensive analysis ("discuss"), as shown in the next table.

Question	Answer guidance
State what is meant by momentum.	mass of an object $\times$ its velocity momentum is a vector quantity
Outline why the kinetic model of an ideal gas is an example of a theoretical model rather than an empirical model.	Empirical implies that the result follows from an experiment. The kinetic model is theoretical and follows from a series of assumptions.
Explain why there is a force acting on a garden hose when water is moving through the hose.	The answer should consider the momentum of the system when water is leaving the hose.
A boy throws a ball vertically into the air and later catches it. *Discuss*, with reference to Newton's third law, the changes in velocity that occur to the ball and the Earth at the instants when the ball is released and caught.	The answer will need to: • state Newton's third law of motion • consider the action-reaction pairs involved • discuss the magnitude and direction of the velocities when the ball is thrown and when it is caught.

A list of commonly used command terms in Physics examination questions is given in the table below. Understanding the exact meaning of frequently used command terms is essential for your success in the examination. Therefore, you should explore this table and use it regularly as a reference when answering questions in this book.

Command term	Definition
Annotate	Add brief notes to a diagram or graph
Calculate	Obtain a numerical answer showing your working
Comment	Give a judgment based on a given statement or result of a calculation
Compare	Give an account of the similarities between two or more items
Compare and contrast	Give an account of similarities and differences between two or more items
Construct	Present information in a diagrammatic or logical form
Deduce	Reach a conclusion from the information given
Describe	Give a detailed account
Determine	Obtain the only possible answer
Discuss	Offer a considered and balanced review that includes a range of arguments, factors or hypotheses
Distinguish	Make clear the differences between two or more items
Draw	Represent by a labelled, accurate diagram or graph, drawn to scale, with plotted points (if appropriate) joined in a straight line or smooth curve

Continued on page VI

Command term	Definition
Estimate	Obtain an approximate value
Explain	Give a detailed account including reasons or causes
Formulate	Express precisely and systematically a concept or argument
Identify	Provide an answer from a number of possibilities
Justify	Give valid reasons or evidence to support an answer or conclusion
Label	Add labels to a diagram
List	Give a sequence of brief answers with no explanation
Outline	Give a brief account or summary
Predict	Give an expected result
Sketch	Represent by means of a diagram or graph (labelled as appropriate), giving a general idea of the required shape or relationship
State	Give a specific name, value or other brief answer without explanation
Suggest	Propose a solution, hypothesis or other possible answer

There is a fuller explanation of all the command terms and additional advice for answering each one online at **www.oxfordsecondary.com/ib-prepared-support**.

Preparation and exam strategies

In addition to the above suggestions, there are some simple rules you should follow during your preparation study and the exam itself.

1. **Get ready for study.** Have enough sleep, eat well, drink plenty of water and reduce your stress by positive thinking and physical exercise. A good night's sleep is particularly important before the exam day, as it can improve your score.

2. **Organize your study environment.** Find a comfortable place with adequate lighting, temperature and ventilation. Avoid distractions. Keep your papers and computer files organized. Bookmark useful online and offline material.

3. **Plan your studies.** Make a list of your tasks and arrange them by importance. Break up large tasks into smaller, more manageable parts. Create an agenda for your studying time and make sure that you can complete each task on time.

4. **Use this book as your first point of reference.** Work your way through the topics systematically and identify any gaps in your understanding and skills. Spend extra time on the topics where improvement is required. Check your textbook and online resources for more information.

5. **Read actively.** Focus on understanding rather than rote learning. Recite key points and definitions using your own words. Try to solve every worked example and practice problem before looking at the answer. Make notes for future reference.

6. **Get ready for the exams.** Practice answering exam-style questions under a time constraint. Learn how to use the Physics data booklet quickly and efficiently. Solve as many problems from past papers as you can. Take a trial exam using the papers at the end of this book.

7. **Optimize your exam approach.** Read all questions carefully, paying extra attention to command terms. Keep your answers as short and clear as possible. Double-check all numerical values and units and quote answers to sensible numbers of significant figures. Label graph axes with the quantity and unit and annotate diagrams. Use exam tips from this book.

8. **Do not panic.** Take a positive attitude and concentrate on things you can improve. Set realistic goals and work systematically to achieve them. Reflect on your performance and learn from mistakes to improve your future results.

Key features of the book

Each chapter typically covers one core or option topic, and starts with "**You should know**" and "**You should be able to**" checklists. These outline the *understandings* and *applications and skills* sections of the IB diploma Physics syllabus. Some assessment statements have been reworded or combined together to make them more accessible and simplify the navigation. These changes do not affect the coverage of key syllabus material, which is always explained within the chapter.

Chapters contain the features outlined on this page.

Theoretical concepts and key definitions are discussed at a level sufficient for answering typical examination questions. Many concepts are illustrated by diagrams, tables or worked examples. Most definitions are explained in a grey side box like this one.

Example

Examples offer solutions to typical problems and demonstrate common problem-solving techniques.

 Nature of science relates a physics concept to the overarching principles of the scientific approach and the development of your own learning skills.

Sample student answers show typical student responses to IB-style questions (most of which are taken from past examination papers). In each response, the correct points are often highlighted in green while incorrect or incomplete answers are highlighted in red. Positive or negative feedback on the student's response is given in the green and red pull-out boxes. An example is given below.

>> **Assessment tip**

This feature highlights essential terms and statements that have appeared in past markschemes, warns against common errors and shows how to optimize your approach to particular questions.

Links provide a reference to relevant material, within another part of this book or the IB Physics data booklet, that relates to the text in question.

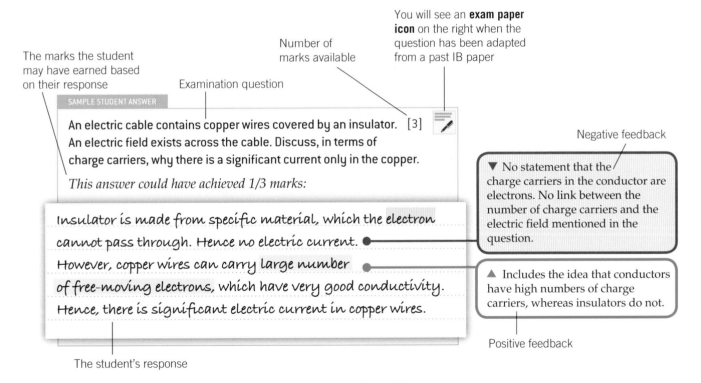

You will see an **exam paper icon** on the right when the question has been adapted from a past IB paper

Number of marks available

The marks the student may have earned based on their response

Examination question

SAMPLE STUDENT ANSWER

An electric cable contains copper wires covered by an insulator. [3]
An electric field exists across the cable. Discuss, in terms of charge carriers, why there is a significant current only in the copper.

This answer could have achieved 1/3 marks:

Insulator is made from specific material, which the electron cannot pass through. Hence no electric current.

However, copper wires can carry large number of free-moving electrons, which have very good conductivity. Hence, there is significant electric current in copper wires.

The student's response

Negative feedback

▼ No statement that the charge carriers in the conductor are electrons. No link between the number of charge carriers and the electric field mentioned in the question.

▲ Includes the idea that conductors have high numbers of charge carriers, whereas insulators do not.

Positive feedback

Questions not taken from past IB examinations will not have the exam paper icon.

Practice problems

Practice problems are given at the end of each chapter. These are IB-style questions that provide you with an opportunity to test yourself and improve your problem-solving skills. Some questions introduce factual or theoretical material from the syllabus that can be studied independently.

1 MEASUREMENTS AND UNCERTAINTIES

1.1 MEASUREMENTS IN PHYSICS

You must know:

✔ the definitions of fundamental and derived SI units

✔ what is meant by scientific notation

✔ the meaning of metric multipliers

✔ that significant figures are used to indicate levels of precision in measurements

✔ what is meant by an order of magnitude

✔ what is meant by an estimation.

You should be able to:

✔ use SI units in a correct format when expressing measurements, final calculated answers and when you are presenting raw and processed data

✔ use scientific notation in conjunction with metric multipliers to express answers and data in as concise a way as possible

✔ quote and compare ratios, values, estimations and approximations to the nearest order of magnitude

✔ estimate quantities to an appropriate number of significant figures.

🔗 The change in definitions of the SI fundamental units in May 2019 does not affect your IB Diploma Programme (DP) learning as you are not required to know the definitions except as indicated in the subject guide. However, you should be aware that textbooks written before this date may give the older definitions.

> **» Assessment tip**
>
> In physics, unless you are providing a final answer as a ratio or as a fractional difference, you must **always** quote the correct unit with your answer. Marks can be lost in an examination when a unit is missing or is incorrect.
>
> You should always link your answer value to its unit (together with the prefix where appropriate).

Scientists need a shared language to communicate between themselves and with the wider public. Part of this language involves agreeing the units used to specify data. For example, if you are told that your journey to school has a value of 5000 then you need to know whether this is measured in metres (originally a European measure) or fet (an old Icelandic length measure).

The agreed set of units and rules is known as the *Système Internationale d'Unités* (almost always abbreviated as SI). In this system, seven *fundamental (base) units* are defined and all other units are derived from these. You are required to use six of the seven fundamental units; the seventh is the unit of luminous intensity, the candela, that is not used in the IB Diploma Programme physics course.

The six fundamental units you will use in the DP physics course are shown in this table.

Measure	Unit	Abbreviation
mass	kilogramme	kg
length	metre	m
time	second	s
quantity of matter	mole	mol
temperature	kelvin	K
current	ampère	A

There are many other derived units used in the course and the expression of these in fundamental units is usually given in this book when you meet the derived unit for the first time. Examples of these derived units include joule, volt, watt, pascal.

Often, the use of a derived unit avoids a long string of fundamental units at the end of a number, so $1 \text{ volt} \equiv 1 \text{ J C}^{-1} \equiv 1 \text{ kg m}^2 \text{ s}^{-3} \text{ A}^{-1}$.

There are also some units used in the course that are not SI. Examples include MeV c^{-2}, light year and parsec. These have special meaning in some parts of the subject and are used by scientists in those fields. Their meaning is explained when you meet them in this book.

The SI also specifies how data in science should be written. Numbers in physics can be very large or very small. Expressing the diameter of an atom as $0.000\,000\,000\,12$ m is unhelpful; 1.2×10^{-10} m is much better. This format of $n.nn \times 10^n$ is known as *scientific notation* and should be used whenever possible. It can also be combined with the SI prefixes that are permitted.

SI prefixes are added in front of a unit to modify its value, so 1012 s can be written as 1.012 ks. The full list of prefixes that you are allowed is included in the data booklet and you can refer to it during examinations.

Prefix	Symbol	Factor	Decimal number
deca	da	10^1	10
hecto	h	10^2	100
kilo	k	10^3	1 000
mega	M	10^6	1 000 000
giga	G	10^9	1 000 000 000
tera	T	10^{12}	1 000 000 000 000
peta	P	10^{15}	1 000 000 000 000 000
deci	d	10^{-1}	0.1
centi	c	10^{-2}	0.01
milli	m	10^{-3}	0.001
micro	μ	10^{-6}	0.000 001
nano	n	10^{-9}	0.000 000 001
pico	p	10^{-12}	0.000 000 000 001
femto	f	10^{-15}	0.000 000 000 000 001

There are some rules here too.

• Only one prefix is allowed per unit, so it would be incorrect to write $2.5\,\mu\text{kg}$ for 2.5 mg.

• You can put one prefix per fundamental unit, so 0.33 Mm ks^{-1} would be acceptable for 330 m s^{-1} (the speed of sound in air) but nowhere near as meaningful.

Significant figures (sf) can lead to confusion. It is important to distinguish between significant figures and decimal places (dp). For example:

• 2.38 kg has 3 sf and 2 dp

• 911.2 kg has 4 sf and 1 dp.

The rule for the number of sf in a calculated answer is quite clear. Specify the answer to the same number as the quantity in the question with the smallest number of sf.

>> **Assessment tip**

Many marks are lost through careless use of units in every DP physics examination. When a question begins 'Calculate, in kg, the mass of…', if you do not quote a unit for your answer then the examiner will assume that you meant kg. If you worked the answer out in g and did not say so, then you will lose marks.

Assessment tip

In Example 1.1.1, rounding up is needed. You should do this for every calculation– *but only at the very end of the calculation*. Rounding answers mid-solution leads to inaccuracies that may take you out of the allowed tolerance for the answer. Keep all possible sf in your calculator until the end and only make a decision about the sf in the last line. In Example 1.1.1, an examiner would be very happy to see …

$= 1.073718 \times 10^{-3}\,m\,s^{-1}$ so the speed of the snail is $1.1 \times 10^{-3}\,m\,s^{-1}$ (to 2 sf) … as your working is then completely clear.

Assessment tip

You may see order of magnitude answers in Paper 1 (multiple choice) written as a single integer. When the response is, say, 7, this will mean 10^7.

It is also permissible to talk about 'a difference of two orders of magnitude'; this means a ratio of 100 (10^2) between the two quantities.

Assessment tip

If the command term 'Estimate' is used in the examination, it will always be clear what is required as you will lack some or all data for your calculation if an educated guess is needed. In estimation questions, such as Example 1.1.2, make it clear what numbers you are providing for each step and how they fit into the overall calculation.

Example 1.1.1

A snail travels a distance of 33.5 cm in 5.2 minutes.

Calculate the speed of the snail.

State the answer to an appropriate number of significant figures.

Solution
The answer, to 7 sf, is $1.073718 \times 10^{-3}\,m\,s^{-1}$.

It is incorrect to quote the answer to this precision as the time is only quoted to 2 sf (the fact that 5.2 minutes is 312 s is not important). The appropriate answer is $1.1 \times 10^{-3}\,m\,s^{-1}$ (or $1.1\,mm\,s^{-1}$ if you prefer).

Sometimes estimations are required in physics. This is because either:

- an educated guess is needed for all or some of the quantities in a calculation, or
- there is an assumption involved in a calculation.

Often it will be appropriate to express your answer to an order of magnitude, meaning rounded to the nearest power of ten. The best way to express any order of magnitude answers is as 10^n, where n is an integer.

Example 1.1.2

Estimate the number of air molecules in a room.

Solution
The calculation is left for you, but you should use the following steps.

- Estimate the volume of a room by making an educated guess at its dimensions, in metres.
- The density of air is about $1.3\,kg\,m^{-3}$—call it $1\,kg\,m^{-3}$ to make the numbers easy later.
- The mass of 1 mol of oxygen molecules is 32 g and 1 mol of nitrogen is 28 g—call the answer 30 g for both gases combined.
- Each mole contains 6×10^{23} molecules.

The volume and density → mass of gas in room and molar mass → number of moles and Avogadro's number → answer.

1.2 UNCERTAINTIES AND ERRORS

You must know:

✓ what is meant by random errors and systematic errors

✓ what is meant by absolute, fractional and percentage uncertainties

✓ that error bars are used on graphs to indicate uncertainties in data

✓ that gradients and intercepts on graphs have uncertainties.

You should be able to:

✓ explain how random and systematic errors can be identified and reduced

✓ collect data that include absolute and/or fractional uncertainties and go on to state these as an uncertainty range

✓ determine the overall uncertainty when data with uncertainties are combined in calculations involving addition, subtraction, multiplication, division and raising to a power

✓ determine the uncertainty in gradients and intercepts of graphs.

All measurement is prone to error. The Heisenberg uncertainty principle (Topic 12) reminds us of the fundamental limits beyond which science cannot go. However, even when the data collected are well above this limit, then two basic types of error are implicit in the data you collect: *random error* and *systematic error*.

Random errors lead to an uncertainty in a value. One way to assess their impact on a measurement is to repeat the measurement several times and then use half the range of the outlying values as an estimate of the *absolute uncertainty*.

> **Random errors** are unpredictable changes in data collected in an experiment. Examples include fluctuations in a measuring instrument or changes in the environmental conditions where the experiment is being carried out.
>
> **Systematic errors** are often produced within measuring instruments. Suppose that an ammeter gives a reading of +0.1 A when there is no current between the meter terminals. This means that every reading made using the meter will read 0.1 A too high. The effect of a systematic error can produce a non-zero intercept on a graph where a line through the origin is expected.

> Uncertainty in measurement is expressed in three ways.
>
> **Absolute uncertainty**: the numerical uncertainty associated with a quantity. For example, when a length of quoted value 5.00 m has an actual value somewhere between 4.95 m and 5.05 m, the absolute uncertainty is ± 0.05 m.
>
> The length will be expressed as (5.00 ± 0.05) m.
>
> **Fractional uncertainty** $= \dfrac{\text{absolute uncertainty in quantity}}{\text{numerical value of quantity}}$.
> A fractional uncertainty has no unit.
>
> **Percentage uncertainty** $=$ fractional uncertainty $\times 100$ expressed as a percentage. There is no unit.

Example 1.2.1

Five readings of the length of a small table are made. The data collected are:

0.972 m, 0.975 m, 0.979 m, 0.981 m, 0.984 m

a) Calculate the average length of the table.

b) Estimate, for the length of the table, its:

    i) absolute uncertainty

    ii) fractional uncertainty

    iii) percentage uncertainty.

Solution

a) The average length is:

$$\frac{(0.972 + 0.975 + 0.979 + 0.981 + 0.984)}{5} = 0.978(2)\,\text{m}$$

b) i) The outliers are 0.972 and 0.984 which differ by 0.012 m. Half this value is 0.006 m and this is taken to be the absolute uncertainty.

The length should be expressed as $(0.978 \pm 0.006)\,\text{m}$.

(This absolute error is an estimate; another estimate is the standard deviation of the set of measurements which in this case is 0.004 m. 0.006 m is thus an overestimate.)

ii) The fractional uncertainty is $\frac{0.006}{0.9782} = 0.006(13) = 0.006$.

This is a ratio of lengths and has no unit.

iii) The percentage uncertainty is $0.006 \times 100 = 0.6\%$.

You will often need to combine quantities mathematically: a pair of lengths, both with uncertainty, may need to be added to give a total length. This derived quantity will also have an uncertainty.

Combining uncertainties

The two sides of a table have lengths (180 ± 5) cm and (60 ± 3) cm. What is the total perimeter of the table?

The **absolute uncertainties are added** when quantities are **added and subtracted**.

When $y = a \pm b$ then $\Delta y = \Delta a + \Delta b$

In this case, the perimeter of the table is $180 + 180 + 60 + 60 = 480$ m. The absolute uncertainty is $5 + 5 + 3 + 3 = 16$ cm.
The perimeter is (480 ± 16) cm or 4.8 ± 0.2 m.

Notice that when the quantities themselves are subtracted, the uncertainties are still added.

What is the area of the table?

When $y = \dfrac{ab}{c}$ then $\dfrac{\Delta y}{y} = \dfrac{\Delta a}{a} + \dfrac{\Delta b}{b} + \dfrac{\Delta c}{c}$

The **fractional uncertainties are added** when quantities are **multiplied or divided**.

The area is $1.8 \times 0.60 = 1.08$ m². The two fractional uncertainties are
$\dfrac{0.05}{1.8} = 0.028$ and $\dfrac{0.03}{0.6} = 0.050$.

The sum is 0.078 and this is the fractional uncertainty of the answer.

The absolute uncertainty in the area $= 0.078 \times 1.08 = 0.084$.

The answer should be expressed as (1.08 ± 0.08) m².

When the answer is found by division, the fractional uncertainties are still added.

Raising quantities to a power

When $y = a^2$, this is the same as $a \times a$ so using the algebraic rule above: $\dfrac{\Delta y}{y} = \dfrac{\Delta a}{a} + \dfrac{\Delta a}{a} = \dfrac{2\Delta a}{a}$.

In the general case, when $y = a^n$, $\dfrac{\Delta y}{y} = \left| n\dfrac{\Delta a}{a} \right|$, where $||$ means the absolute value or magnitude of the expression.

When **a quantity is raised to a power n, the fractional uncertainty is multiplied by n**.

The radius of a sphere is (0.20 ± 0.01) m. What is the volume of the sphere?

Volume of sphere is: $\dfrac{4}{3}\pi r^3 = 0.0335$ m³

where r is the radius.

Fractional uncertainty of radius $= \dfrac{0.01}{0.20} = 0.05$

So, the fractional uncertainty of the radius cubed is $3 \times 0.05 = 0.15$.

The absolute uncertainty is $0.335 \times 0.15 = 0.0050$ m³.

The volume of the sphere is (0.335 ± 0.005) m³.

There is more information about this topic in Chapter 13, which deals with Paper 3, Section A.

It is possible that data points, all with an associated error, are presented on a graph. Therefore, there are errors associated with the gradient and any intercept on the graph. The way to treat these errors is to add *error bars* to the graph. These are vertical or horizontal lines, centred on each data point, that are equal to the length of the absolute errors.

Maximum and minimum best-fit lines can then be drawn each side of the true best-fit line. The gradients of these maximum–minimum lines give a range of values that corresponds to the error in the gradient. The intercepts of the maximum–minimum lines also have a range in values that can be associated with the error in the true intercept.

For the graph in Figure 1.2.1, the gradient is 1.6 with a range between 2.1 and 1.1, so $(1.6 \pm 0.5)\,\mathrm{m\,s^{-1}}$.

The intercept is −2.4 with a range of 1.0 to −5.8, so $(-2.4 \pm 3.4)\,\mathrm{m}$.

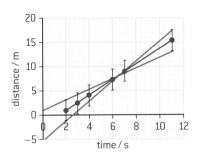

▲ Figure 1.2.1. Maximum and minimum best-fit lines each side of a true best-fit line

1.3 VECTORS AND SCALARS

You must know:

✔ what are meant by vector and scalar quantities

✔ that vectors can be combined and resolved (split into two separate vectors).

You should be able to:

✔ solve vector problems graphically and algebraically.

Quantities in DP physics are either *scalars* or *vectors*. (There is a third type of physical quantity but this is not used at this level.)

A vector can be represented by a line with an arrow. When drawn to scale, the length of the line represents the magnitude, and the direction is as drawn.

Both scalars and vectors can be added and subtracted. Scalar quantities add just as any other number in mathematics. With vectors, however, you need to take the direction into account.

Figure 1.3.1 shows the addition of two vectors. The vectors must be drawn to the same scale and the direction angles drawn accurately too. A further construction produces the parallelogram with the red solid and dashed lines. Then the magnitude of the new vector $\mathbf{v}_1 + \mathbf{v}_2$ is given by the length of the blue vector with the direction as shown.

> **Scalars** are quantities that have magnitude (size) but no direction. They generally have a unit associated with them.
>
> **Vectors** are quantities that have both magnitude and a physical direction. A unit is associated with the number part of the vector.
>
> For example, the scalar quantity speed is written as v; the vector quantity velocity is written as $\mathbf{v}$ (sometimes as $\underline{v}$ or $\vec{v}$, but this notation is not used in this book).

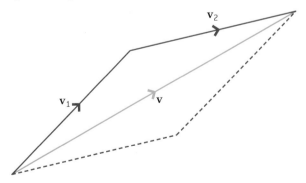

▲ Figure 1.3.1. Adding vectors $\mathbf{v}_1$ and $\mathbf{v}_2$

Vectors can also be added algebraically. The most common situation you meet in the DP physics course is when the vectors are at 90° to each other (Figure 1.3.2).

As before, addition by drawing gives the red vector which is the sum of $\mathbf{v}_1$ and $\mathbf{v}_2$. Algebraically, the use of trigonometry gives the magnitude of the resultant (added) vector as $\sqrt{\mathbf{v}_1^2 + \mathbf{v}_2^2}$ and the direction θ as $\tan^{-1}\left(\dfrac{\mathbf{v}_2}{\mathbf{v}_1}\right)$.

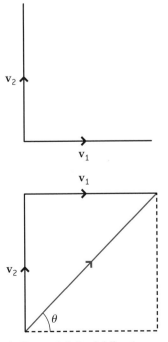

▲ Figure 1.3.2. Adding two vectors at right angles

Example 1.3.1

A girl walks 500 m due north and then 1200 m due east. Calculate her position relative to her starting point.

Solution

This is similar to the situation in Figure 1.3.2 where the first vector has a magnitude of 500 m and the second a magnitude of 1200 m.

The magnitude of the resultant is $\sqrt{500^2 + 1200^2} = 1300$ m.

θ is $\tan^{-1}\left(\dfrac{500}{1200}\right) = 22.6°$.

Another skill required in the DP physics course is that of breaking a vector down into two components at right angles to each other – this is known as resolving the vector. A right angle is chosen because the two resolved components will be independent of each other. Figure 1.3.3 shows the process.

The vector **F** points upwards from the horizontal at θ. This length **F** is the hypotenuse of the right-angled triangle. The other sides have lengths **F**$\cos\theta$ and **F**$\sin\theta$.

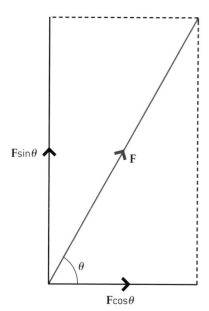

▲ Figure 1.3.3. Resolving a vector

Example 1.3.2

An object moves with a velocity 40 m s⁻¹ at an angle N30°E. Determine the component of the velocity in the direction:

a) due east

b) due north.

Solution

a) The angle between the vector and east is 60°

So the component due east = $40\cos 60° = 20\,\mathrm{m\,s^{-1}}$

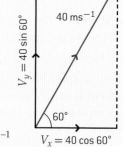

b) Due north, the component is $40\cos 30° = 40\sin 60° = 34.6\,\mathrm{m\,s^{-1}}$

Example 1.3.3

A girl cycles 1500 m due north, 800 m due east and 1000 m in a south-easterly direction. Calculate her overall displacement.

Solution

A drawing of the journey is shown. The total horizontal component of the displacement is $800 + 1000\cos 45° = 1510$ m. The total vertical component is $1500 - 1000\cos 45° = 790$ m.

The displacement is 1700 m at $\tan^{-1}\left(\dfrac{790}{1510}\right) = 28°$.

You can now add or subtract any non-parallel vectors algebraically. Figure 1.3.4 shows the method.

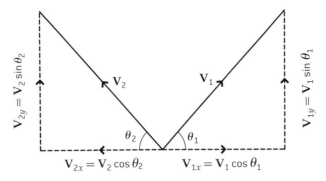

▲ Figure 1.3.4. Algebraic method for adding or subtracting non-parallel vectors

Horizontally the addition gives $\mathbf{V}_x = \mathbf{V}_{1x} + \mathbf{V}_{2x}$ which is $\mathbf{V}_1 \cos\theta_1 - \mathbf{V}_2 \cos\theta_2$.

Vertically the addition gives $\mathbf{V}_y = \mathbf{V}_{1y} + \mathbf{V}_{2y}$ which is $\mathbf{V}_1 \sin\theta_1 + \mathbf{V}_2 \sin\theta_2$. These new vector lengths can be added to give the new vector length $\mathbf{V} = \sqrt{\mathbf{V}_x^2 + \mathbf{V}_y^2}$ with an angle to the horizontal of $\tan^{-1} = \left(\dfrac{\mathbf{V}_y}{\mathbf{V}_x}\right)$.

To subtract two vectors, simply form the negative vector of the one being subtracted (by reversing its original direction but leaving the length unchanged) and add this to the other vector.

Practice problems for Topic 1

Problem 1
You will need to have covered the relevant topic before answering this question.

a) Express the following derived units in fundamental units: watt, newton, pascal, tesla.

b) Give a suitable set of fundamental units for the following quantities:
acceleration, gravitational field strength, electric field strength, energy.

Problem 2
Express the following physical constants (all in the data booklet) to the specific number of significant figures.

Quantity	Significant figures required
Neutron rest mass	3
Planck's constant	2
Coulomb constant	2
Permeability of free space	5

Problem 3
Express the following numbers in scientific notation to three significant figures.

a) 4903.5

b) 0.005194

c) 39.782

d) 9273844.45

e) 0.035163

Problem 4
Estimate these quantities.

a) Length of a DP physics course in seconds.

b) Number of free electrons in the charger lead to your computer.

c) Volume of a door.

d) Number of atoms in a chicken's egg (assume it is made of water).

e) Number of molecules of ink in a pen.

f) Energy stored in an AA cell.

g) Number of seconds you have been alive.

h) Thickness of tread worn off a car tyre when it travels 10 km.

Problem 5
Determine, the following, with their absolute and percentage uncertainties.

a) The kinetic energy of a mass (1.5 ± 0.2) kg moving at (21.5 ± 0.3) m s^{-1} (use $E_k = \frac{1}{2} mv^2$).

b) The force acting on a wire of length (3.5 ± 0.4) m carrying a current (2.5 ± 0.2) A in a magnetic field of strength (5.2 ± 0.3) mT (use $F = BIL$).

c) The quantity of gas, in mol, in a gas of volume (1.25 ± 0.03) m^3, pressure $(2.3 \pm 0.1) \times 10^5$ Pa at a temperature of (300 ± 10) K (use $pV = nRT$).

Problem 6
A car is driven at 30 m s^{-1} for 30 minutes due east and then at 25 m s^{-1} for 45 minutes northeast.

Calculate the final displacement of the car from its starting point.

2 MECHANICS

2.1 MOTION

You must know:

✔ the meaning of the terms distance, displacement, speed, velocity and acceleration

✔ the difference between instantaneous and average values for velocity, speed and acceleration

✔ that the kinematic equations of motion apply only to uniform acceleration

✔ how fluid resistance affects motion

✔ what is meant by terminal speed and fluid resistance.

You should be able to:

✔ use the kinematic equations of motion (*suvat*) for uniform acceleration

✔ represent and interpret motion using distance–time, speed–time, displacement–time, velocity–time and acceleration–time graphs

✔ resolve acceleration, velocity and displacement into vertical and horizontal components

✔ analyse projectile motion

✔ describe an experimental determination of the acceleration of free-fall.

🔗 Vector and scalar quantities are discussed in Topic 1.3 Vectors and scalars.

> **» Assessment tip**
>
> When a question asks for a vector quantity, such as velocity or displacement, you must give the direction **and** magnitude of the quantity to gain full marks.

Distance is a scalar quantity: it is the length of a path between two points.

Speed is a scalar quantity:

$$\text{average speed} = \frac{\text{total distance travelled by an object}}{\text{total time taken}}.$$

Displacement is a vector quantity: it is the difference between an initial position and a final position.

Velocity is a vector quantity:

$$\text{average velocity} = \frac{\text{change in displacement}}{\text{time taken for change}}.$$

Acceleration is $\dfrac{\text{change in velocity}}{\text{time taken for change}}$.

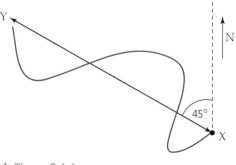

▲ Figure 2.1.1.

The motion of an object is described by the vector quantities *displacement*, *velocity* and *acceleration*. The scalar counterparts of these are *distance* and *speed*.

Look at Figure 2.1.1. The length of the curved path is the **distance**. The direct line from X to Y is the **displacement**. Both the length of XY and its direction relative to north are required to specify the displacement completely.

Motion can be represented using distance–time and speed–time graphs. Vector quantities can be shown as displacement–time, velocity–time and acceleration–time graphs.

Example 2.1.1

A ball is released, from rest, from a distance above the ground. This graph plots *vertical speed of the ball v* against *time t*.

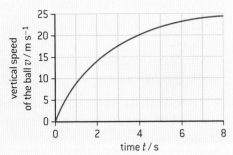

a) Estimate the distance fallen by the ball from $t = 0$ to $t = 8.0$ s.

b) Determine the acceleration of the ball when $t = 4.0$ s.

Solution

a) The distance fallen by the ball is equal to the area under the curve. To calculate this area, count the number of whole squares under the curve. Then multiply this number by the area of one square.

In this example, there are approximately 28 squares under the curve. Each square has an area of $1\,\text{s} \times 5\,\text{m s}^{-1}$, which is equivalent to a distance of $1 \times 5 = 5\,\text{m}$.

So the distance travelled is $28 \times 5 = 140\,\text{m}$.

b) You can find the acceleration of the ball when $t = 4.0$ s by drawing a tangent to the curve at $t = 4.0$ s. The gradient of this tangent is the acceleration. Make sure you read off values as far apart as possible (at least half the length of the line).

$$\text{acceleration} = \frac{25 - 12}{6.4} = 2.03 \equiv 2.0\,\text{m s}^{-2}$$

The *suvat* equations apply for uniform acceleration **only**; 'uniform' means 'constant' here. The speed–time graph must be a straight line.

Example 2.1.2

A car changes its speed uniformly from $28\,\text{m s}^{-1}$ to $12\,\text{m s}^{-1}$ in 8.0 s.

a) Calculate the acceleration of the car.

b) Determine the distance travelled by the car during the 8.0 s of acceleration.

Solution

a) Use the *suvat* equations to calculate the acceleration. Start by writing down what you know, including symbols and units, and what you want to find out.

u	initial speed	$28\,\text{m s}^{-1}$
v	final speed	$12\,\text{m s}^{-1}$
a	acceleration	$?\,\text{m s}^{-2}$
t	change in time	8.0 s

The equation you need is $v = u + at$.

Note that the units match; if they didn't, you would need to convert them. Take care to substitute correctly into the equation:

$12 = 28 + a \times 8.0$ and therefore $a = -2.0\,\text{m s}^{-2}$

The car is slowing down so the acceleration is negative.

b) Use $s = ut + \frac{1}{2}at^2 = (28 \times 8) + \frac{1}{2} \times -2.0 \times 8^2 = 224 - 64 = 160\,\text{m}$

There is more advice for constructing graphs and determining the gradient of a graph at a point and the area under the graph in Chapter 13.

>> **Assessment tip**

Quantities can be derived from the motion graphs as follows:

- Distance–time graph: **speed** is the **gradient** of the graph
- Speed–time graph: **distance travelled** is the **area** under the graph and **acceleration** is the **gradient** of the graph
- Acceleration–time graph: **speed change** is the **area** under the graph.

The symbols used for motion in the DP physics course are:

s	displacement
u	initial speed
v	final speed
a	acceleration
t	change in time

These are connected by the four kinematic equations of motion for uniform acceleration, often known as the ***suvat*** equations:

$$v = u + at$$
$$v^2 = u^2 + 2as$$
$$s = ut + \frac{1}{2}at^2$$
$$s = \frac{(v + u)t}{2}$$

The kinematic equations for uniform acceleration can be used to determine the acceleration of free fall g for objects released near the Earth's surface; the distance fallen from rest and time to fall are required.

> **Assessment tip**

Look for sentences such as 'Assume air resistance is negligible'; this makes your job easier. You will not have to carry out calculations involving air resistance for projectile motion, but you may have to answer qualitative (non-mathematical) questions about it.

The resolution of a vector into two components at right angles is considered in Topic 1.3.

Solving questions involving *projectile motion* involves:

- resolving the initial velocity into horizontal and vertical components
- recognizing that the acceleration due to free-fall acts only on the vertical component.

As g is constant, the *suvat* equations can be used. The horizontal component of velocity does **not** change (assuming negligible air resistance).

Example 2.1.3

Idris kicks a soccer ball over a wall that is a horizontal distance of 32 m away. Assume air resistance is negligible.

a) The ball takes 1.6 s to reach a point vertically above the wall. Calculate the horizontal component of the velocity of the ball.

b) The ball is at its maximum height as it passes above the wall. State the vertical component of its velocity at maximum height.

c) Determine the magnitude of the initial velocity of the ball when it was kicked.

d) Deduce the angle to the horizontal θ at which the ball was kicked.

Solution

a) The ball travels 32 m horizontally in a time of 1.6 s.

The horizontal component of velocity is $\dfrac{32}{1.6} = 20.0 \, \text{m s}^{-1}$.

b) At its maximum height, the ball moves horizontally only, so the vertical component is zero.

c) You need to know the initial vertical component of the velocity, so you can combine it with the horizontal component. Use a *suvat* equation.

u	initial speed	? m s^{-1}
v	final speed	0
t	change in time	1.6 s

(Note that acceleration a is negative because the initial component upwards is positive.)

Use $v = u + at$. Substituting gives $0 = u + -9.8 \times 1.6$

Rearranging gives $u = 15.68 \, \text{m s}^{-1}$.

You now have the information needed to calculate the vector velocity. Its magnitude is: $\sqrt{20^2 + 15.68^2} = 25.4 \, \text{m s}^{-1} = 25 \, \text{m s}^{-1}$ (2 s.f.).

d) The angle to the horizontal is $\tan^{-1} \dfrac{15.68}{20} = 38°$.

The effect of air resistance on an object's motion is covered in more detail in Topic 2.2.

Resistive forces oppose the motion of objects moving in a gas or in a liquid. They are considered on page 15.

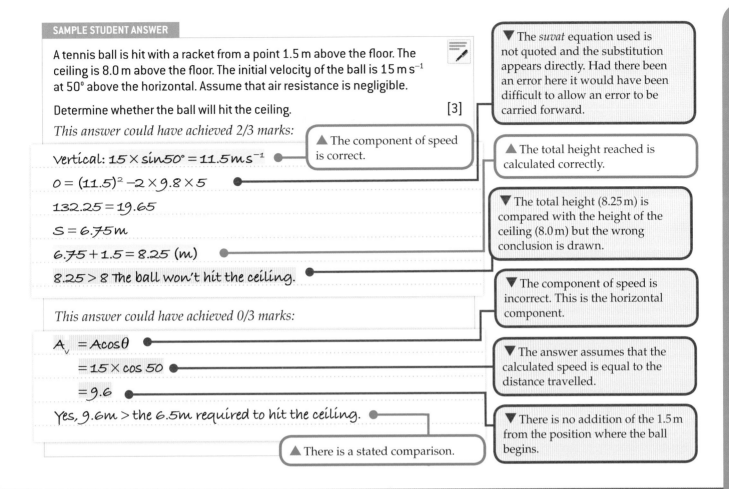

SAMPLE STUDENT ANSWER

A tennis ball is hit with a racket from a point 1.5 m above the floor. The ceiling is 8.0 m above the floor. The initial velocity of the ball is 15 m s⁻¹ at 50° above the horizontal. Assume that air resistance is negligible.

Determine whether the ball will hit the ceiling. [3]

This answer could have achieved 2/3 marks:

Vertical: $15 \times \sin 50° = 11.5 \, m\,s^{-1}$

$0 = (11.5)^2 - 2 \times 9.8 \times 5$

$132.25 = 19.65$

$S = 6.75m$

$6.75 + 1.5 = 8.25 \, (m)$

$8.25 > 8$ The ball won't hit the ceiling.

▲ The component of speed is correct.

▼ The *suvat* equation used is not quoted and the substitution appears directly. Had there been an error here it would have been difficult to allow an error to be carried forward.

▲ The total height reached is calculated correctly.

▼ The total height (8.25 m) is compared with the height of the ceiling (8.0 m) but the wrong conclusion is drawn.

This answer could have achieved 0/3 marks:

$A_v = A\cos\theta$

$= 15 \times \cos 50$

$= 9.6$

Yes, 9.6m > the 6.5m required to hit the ceiling.

▼ The component of speed is incorrect. This is the horizontal component.

▼ The answer assumes that the calculated speed is equal to the distance travelled.

▲ There is a stated comparison.

▼ There is no addition of the 1.5 m from the position where the ball begins.

2.2 FORCES

You must know:

✔ that real objects can be represented by points in space

✔ what is meant by translational equilibrium

✔ Newton's three laws of motion

✔ that the weight of an object is equal to its *mass × acceleration of free fall*

✔ that solid friction can be described using static and dynamic coefficients of friction.

You should be able to:

✔ represent a force as a vector

✔ draw and interpret a free-body diagram

✔ interpret the forces acting on a system in terms of Newton's first law

✔ carry out calculations using Newton's second law

✔ identify action and reaction pairs using Newton's third law.

Forces are said to be pushes or pulls. This implies a contact between one object and another. But forces also arise through field interactions, for example, a mass interacts with a gravitational field. Newton's three laws of motion describe the interaction between an object and a force.

Uniform velocity has both magnitude **and** direction, so **both** must be unchanging for the motion to be uniform.

For acceleration to occur an object must be subject to a net *external force*. When more than one external force acts on a body, you need to sum the vectors using the rules for vector addition. When the forces cancel out to give zero *resultant force (net force)* the body is in *translational equilibrium*. When the forces do not cancel out to give zero resultant force, the forces are *unbalanced*.

Physics often simplifies situations to make the calculations (mathematics) more manageable. It represents real objects using single points. For *translational motion*, this is the centre of mass—where all the mass acts. In gravitational fields, this is also the centre of gravity—where all the weight acts.

>> **Assessment tip**

This book uses N1, N2 and N3 to abbreviate Newton's first, second and third laws of motion. It is best not to use your own abbreviations in examinations unless you state what they are.

Newton's first law of motion (N1) states that an object will remain at rest or continue with uniform velocity unless a net external force acts on it.

Newton's second law of motion (N2) relates the resultant force F acting on an object of mass m to the acceleration of the object a with the equation $F = ma$. Both force and acceleration are vectors, whereas mass is scalar. The direction of the net force and the direction of the acceleration it produces are identical. N2 defines the SI unit of force, the newton. One newton (1 N) is the force that will give a mass of 1 kg an acceleration of $1 \, \text{m s}^{-2}$. In fundamental units, this is kg m s^{-2}.

Newton's third law (N3) states that action and reaction are equal and opposite.

Example 2.2.1

A cyclist travels along a horizontal road at a constant velocity. Explain why the cycle is travelling at constant velocity.

Solution

The system (cyclist and cycle) is moving horizontally at constant velocity. N1 states that, because the velocity is constant, there is no net external force. The forward force of the cyclist pedalling the cycle equals the frictional forces acting in the opposite direction. These forces are equal in magnitude and in opposite directions.

Example 2.2.2

A car of mass $9.0 \times 10^2 \, \text{kg}$ travels in a straight line away from traffic lights after they turn green.

a) Calculate the acceleration of the car between $t = 2.0 \, \text{s}$ and $t = 5.0 \, \text{s}$.

b) Calculate the force used to accelerate the car.

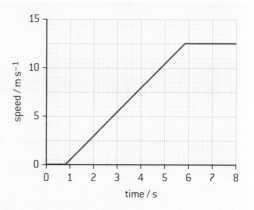

Solution

a) The change in speed is from 3.0 to $10.5 \, \text{m s}^{-1}$. It takes $3.0 \, \text{s}$.

$$\text{acceleration} = \frac{\text{change in speed}}{\text{time taken for change}} = \frac{7.5}{3.0} = 2.5 \, \text{m s}^{-2}$$

b) The force that accelerates the car = mass of car × acceleration
$$= 900 \times 2.5 = 2.25 \, \text{kN}.$$

Since the data used in the calculation were given to 2 significant figures, you should give the answer to 2 significant figures as well (= 2.3 kN).

>> **Assessment tip**

When you **apply** a law in an answer, always **state** the law; that way, the examiner will be clear that you understand the physics underlying your answer.

N3 implies that action–reaction pairs exist. These are pairs of forces that are:

- equal in magnitude and opposite in direction

- the same **type** of force (such as gravitational, electrostatic or tension forces).

When a ball is released from rest above the Earth's surface, a gravitational force acts downwards on the ball due to the attraction of the ball by the Earth. At the same time, the ball attracts the Earth with an upwards gravitational force that is equal in magnitude to the weight of the ball. This is an action–reaction pair.

When the ball sits at rest on a table, an additional force pair now acts. This pair is electrostatic in origin; it is:

- the upwards force of the table on the ball

- the equal downwards force of the ball on the table.

These forces arise because the table and the ball are both deformed by the other object. The electrostatic forces arise from the attempt by each to return to its original shape.

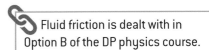
Other interpretations of N2 and N3 are considered in Topic 2.4.

Example 2.2.3

A car of mass 750 kg accelerates at $0.30 \, \text{m s}^{-2}$ in a straight line along a horizontal road. A resistive force of 550 N acts on the car.

a) Calculate the resultant force that acts on the car

b) Calculate the force that the engine exerts on the car.

Solution

a) The accelerating force is $750 \times 0.3 = 225 \, \text{N}$. This is the net force acting. The answer is 230 N, to 2 significant figures.

b) The resistive force is 550 N. The engine must overcome this **and** provide the 230 N. So, the total engine force must be:
$550 + 225 = 775 \, \text{N} \equiv 780 \, \text{N}$.
Note: the full number of significant figures is used until the **final** round-off and answer.

The magnitude of the frictional force acting between two surfaces depends on whether or not the surfaces are moving relative to each other. When there is no relative motion, the friction between the surfaces is called the *static friction*. When there is relative movement, the friction is said to be *dynamic*.

In both cases the magnitude of the frictional force F depends on the normal (perpendicular) reaction force R that acts between the surfaces.

Static friction	Dynamic friction
$F \leq \mu_s R$	$F = \mu_d R$
where μ_s is the coefficient of static friction.	where μ_d is the coefficient of dynamic friction.

In general, the magnitude of static friction is greater than that for dynamic friction, so the size of the frictional force between two surfaces decreases as the surfaces begin to slide.

Different language is used for fluid friction (which occurs in liquids and gases). The terms *air resistance* and *liquid drag* are used to describe the friction on a solid as it moves through a fluid.

Fluid friction is dealt with in Option B of the DP physics course.

The amount of frictional drag between the object and the fluid in which it moves (called *the medium*) depends on the relative speed between them. The higher this relative speed, the greater the drag force.

As an object accelerates, frictional drag increases and the magnitude of the drag eventually equals the accelerating force. The two forces oppose each other, so the net resultant force falls to zero and there is no further change in velocity (N1). The speed at which this occurs is called the *terminal speed*; it depends on the size and shape of the object and the 'stickiness' of the medium. In an automobile, when the maximum force of the engine equals the drag force, the vehicle travels at its terminal speed. Objects that are falling freely in a planet's gravitational field also reach a terminal speed depending on their size and shape.

Example 2.2.4

A small steel ball is released from rest into a fluid. Ignore buoyancy effects in this question.

The graph shows how the speed v of the ball varies with time t.

a) Draw a free-body diagram for the ball at:

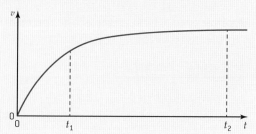

i) time t_1

ii) time t_2.

b) When $t = t_1$, the gradient of the graph is a.

Deduce an expression in terms of the mass of the ball M, acceleration a and acceleration of free fall g for the magnitude of the frictional force F_f acting on the ball at $t = t_1$.

c) Deduce, using your answer to part a) ii), the magnitude of the frictional force acting on the ball when $t = t_2$.

Solution

a) i) The ball is accelerating; there must be a net downward force on it. There is a smaller drag force upwards.

ii) The vector lengths should be equal as the velocity is constant. The net force must be zero (N1).

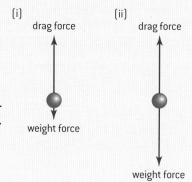

b) The net force $F_{total} = F_{weight} - F_f$,

$F_{weight} = Mg$ and $F_{total} = Ma$

So $F_f = M(g - a)$

c) As the velocity is constant, N1 predicts that the upward and downward forces must be equal. The magnitude of the frictional force (drag) must equal the magnitude of the weight force.

SAMPLE STUDENT ANSWER

An unpowered glider moves horizontally at constant speed. The wings of the glider provide a lift force. The diagram shows the lift force acting on the glider and the direction of motion of the glider.

a) Draw the forces acting on the glider to complete the free-body diagram. The dotted lines show the horizontal and vertical directions.   [2]

This answer could have achieved 2/2 marks:

▲ The diagram is well drawn and everything is clear. The drag opposes the forward motion of the glider and the weight acts downwards, as expected.

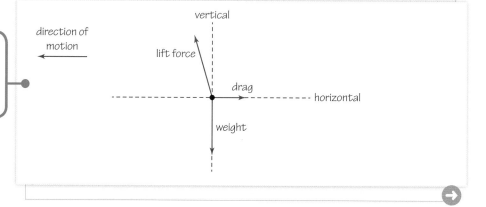

This answer could have achieved 1/2 marks:

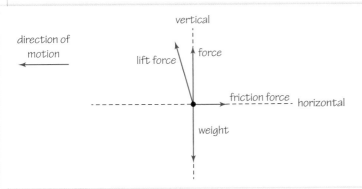

▲ The drag force (here labelled 'friction force') is correct.

▼ However, an upward force is shown acting on the glider. There is certainly a force vertically upwards, but it is already there – it is the vertical component of the lift force.

b) Explain, using appropriate laws of motion, how the forces acting on the glider maintain it in level flight. [2]

This answer could have achieved 2/2 marks:

The 1st Law states that an object will move at a constant speed in a straight line or at rest providing there is no resultant force. The lift's vertical component is the same but opposite to the weight of the glider, meaning there is no net resultant vertical force keeping the glider in level flight.

▲ The law in use is clearly stated (although not attributed to Newton). The example here is referenced clearly and the equality between the vertical component of the lift and the weight is correctly stated.

This answer could have achieved 1/2 marks:

According to Newton's third law, the force A exerting on B is equal to the force B exerting on A. And also the first law the action remains no change or constant when no external force added, hence it maintain its level flight.

▼ There is no link between the quoted laws and the physics of the glider situation. The reference to N3 is not required.

≫ Assessment tip

Do not simply repeat the question back to the examiner. This does not gain credit.

2.3 WORK, ENERGY AND POWER

You must know:

✔ work done is a measure of the energy transferred between two energy stores

✔ there are many forms of energy store, including kinetic, gravitational and elastic

✔ energy can be conserved

✔ power is the rate of energy transfer

✔ efficiency is:

$$\frac{\text{useful work out}}{\text{total work in}} \text{ or } \frac{\text{useful power out}}{\text{total power in}}.$$

You should be able to:

✔ identify the energy stores and pathways for a specified energy transfer

✔ calculate *work done* as *force × distance* and *power* as *force × speed*

✔ interpret force–distance graphs

✔ include the effects of resistive forces in energy-change calculations

✔ calculate changes in kinetic energy and changes in gravitational potential energy

✔ calculate efficiency.

Energy is transferred from one *energy store* to another via an *energy pathway*.

Energy transfers and pathways appear throughout the DP physics course. For example:

• electrical in Topic 5

• heating in Topics 3 and Option B

• waves in Topics 4 and 9.

>> **Assessment tip**

Conservation laws are important in science. They recur many times in physics and are a way to learn (and later revise) the subject effectively. Always look for the links between topics in this subject. Try to link ideas to check your understanding of the whole subject and to ensure that you can think in an unfamiliar context.

---

work done = force (F) × distance (s)

When there is an angle θ between the force direction and the displacement then
work done = $Fs \cos \theta$.

The unit of energy is the **joule** (J). 1 J is equivalent to 1 newton metre (Nm). In fundamental units, this is $\text{kg m}^2\,\text{s}^{-2}$.

Change in kinetic energy ΔE_k of an object of mass m is $\frac{1}{2}m(v^2 - u^2)$ when its speed changes from u to v.

Change in gravitational potential energy ΔE_p of an object of mass m is mgh when it is raised through a vertical height Δh.

>> **Assessment tip**

Conversion of ΔE_k to ΔE_p is a convenient way to solve problems when bodies fall in gravitational fields. You **must** use this approach when the acceleration is not uniform: for example, when a skier slides down a slope with a changing gradient.

Energy stores include:

• elastic and magnetic

• chemical

• kinetic

• gravitational

• electromagnetic

• nuclear

• thermal

Energy pathways (or transfer mechanisms) include:

• electric (a charge moving through a potential difference)

• mechanical (a force acting through a distance)

• heating (driven by a temperature difference)

• waves (such as electromagnetic radiation or sound waves).

The rule of *conservation of energy* is never broken—but you will sometimes have to look hard to see where some energy goes.

In mechanical systems, the energy transferred between stores (such as gravitational to kinetic) is equivalent to the work done.

The equation *work done = force × distance* applies when the force is constant, However when the force is not constant with distance you need to calculate the area under the force distance graph.

Example 2.3.1

A boy drags a box to the right across a rough, horizontal surface using a rope that pulls upwards at 25° to the horizontal.

a) Once the load is moving at a steady speed, the average horizontal frictional force acting on the load is 470 N.

Calculate the average value of F.

b) The load is moved a horizontal distance of 250 m in 320 s. Calculate the work done on the load by F.

Solution

a) The frictional force acts to the left. The horizontal force to the right must equal 470 N for the resultant force to be zero with no change in velocity.

Resolving $F \cos 25 = 470$ gives $F = 518\,\text{N}$ (= 520 N to 2 sf, as for the given data).

b) The work done = force × distance = $518 \times 250 = 130\,\text{kJ}$.

When an object is moving, energy has been transferred into *kinetic energy*. Movement of an object within a gravitational field can also lead to transfer of energy. In this case, the term *gravitational potential energy* is used.

Example 2.3.2

A diver climbs to a diving board and dives from it.

The height of the diving board above the floor is 4.0 m. The mass of the diver is 54 kg.

a) Calculate the gain in gravitational potential energy when the diver climbs to the diving board.

b) The diver enters the water at a speed of 8.0 m s⁻¹. Calculate the kinetic energy of the diver as she enters the water.

c) Suggest why the kinetic energy of the diver in part b) is different from the gravitational potential energy gained in part a).

Solution

a) $\Delta E_k = mgh = 54 \times 9.81 \times 4.0 = 2.1$ kJ

b) $\Delta E_k = \frac{1}{2}m\left(v^2 - u^2\right)$.

Here $u = 0$ as the diver starts from rest.

$\Delta E_k = \frac{1}{2}mv^2 = \frac{1}{2} \times 54 \times 8^2 = 1.7$ kJ

c) A number of factors can be discussed here.

- Work is done against air resistance.

- The distance travelled by the centre of mass of the diver falling to the water is not equal to the distance gained in climbing the stairs.

- The diver gains gravitational potential energy in taking off.

- Energy usually goes into rotational kinetic energy.

Energy can be transferred at different rates. Think of two boys of equal weight who run up a hill: the quicker boy is the more powerful of the two.

As *work done = force × distance*, power $= \dfrac{force \times distance}{time}$ or $force \times \dfrac{distance}{time}$.

Therefore *power = force × speed*. The area under a force–speed graph gives the power developed during an energy transfer.

As you saw in Example 2.3.2, not all energy is necessarily transferred from the original source into its final useful form—in most real cases, there is a transfer via friction. A measure of the effectiveness of a transfer is *efficiency*. This can be defined in terms of energy or in terms of power.

> **Power** is the rate of doing work.
>
> $$power = \frac{energy\ transferred}{time\ taken\ for\ transfer}$$
>
> Its unit is J s⁻¹. 1 J s⁻¹ = 1 **watt** (W). In fundamental units, this is kg m² s⁻³.
>
> $$efficiency = \frac{useful\ energy\ transferred}{total\ energy\ input}$$
>
> $$= \frac{power\ output}{power\ input}$$

Example 2.3.3

Water in a hydroelectric system falls vertically to a river below at the rate of 12 000 kg every minute.

The water takes 2.0 s to fall this distance. It has zero velocity at the top.

a) Calculate the height through which the water falls.

b) A small electrical generator of efficiency 20% is at the foot of the system. All the water goes through the generator.

Determine the electrical power output of the generator.

c) Outline the energy transfers in this system.

19

Solution

a) The acceleration, g, is uniform, so *suvat* equations can be used.

$$s = ut + \frac{1}{2}at^2 = 0 + \frac{1}{2} \times 9.81 \times 2.0^2 = 20\,\text{m to 2 significant figures}$$

b) From this point, it is best to work in seconds rather than minutes.

$$\text{Mass of water flowing every second} = \frac{12\,000}{60} = 200\,\text{kg}$$

Gravitational potential energy transferred every second
$$= mgh = 200 \times 9.81 \times 19.8 = 38\,847.6\,\text{J}$$

As the generator is 20%, the power output is

$$\frac{38\,847.6}{5} = 7769.52\,\text{W} (= 7800\,\text{W to 2 significant figures}).$$

c) The water has stored gravitational potential energy at the top of the waterfall. As the water falls, energy is transferred into kinetic energy. At the bottom of the waterfall, the maximum transfer of energy has occurred. The water enters a turbine where the linear kinetic energy of the water is transferred into rotational kinetic energy of the turbine and the dynamo. The dynamo converts this kinetic energy into electrical energy, wasted thermal energy and frictional losses.

SAMPLE STUDENT ANSWER

An electric motor pulls a glider horizontally from rest to a constant speed of $27.0\,\text{m s}^{-1}$ with a force of 1370 N in a time of 11.0 s. The motor has an overall efficiency of 23.0%.

Determine the average power input to the motor. State your answer to an appropriate number of significant figures.   [4]

This answer could have achieved 3/4 marks:

$$W_{useful} = F.S = 1367.6\,N \times 148.5\,m = 203094\,J$$
$$W_{used} = W_{useful} \times 23\% = 883017.39\,J$$
$$P = \frac{W_{used}}{t} = \frac{W_{used}}{11s} \approx 80274.3\,W$$

▼ This is not the easiest way to carry out the problem. First, the energy gained by the glider is calculated and then the efficiency is used to calculate the energy input to the motor. This is then divided by time to give the power input. There are too many significant figures in the answer which should have been restricted to 2 or 3 given the data in the question.

This answer could have achieved 4/4 marks:

$$\text{Average speed} = \frac{148.5}{11} = 13.5\,ms^{-1}$$
$$\text{Power} = \text{Force} \times \text{velocity} = 1370 \times 13.5 = 18495\,W = 23\%$$
$$\frac{18495}{23} \times 100 = 80413.043\,W$$
$$= 80.4\,kW$$

▲ The solution uses *power = force × speed* leading to the power of the glider directly. Then it uses a simple efficiency calculation and correct rounding to get the answer with an appropriate number of significant figures.

2.4 MOMENTUM

You must know:

✔ that an alternative form for Newton's second law of motion is *force = rate of change of momentum*

✔ that impulse is equal to the change in momentum and this is the area under a force–time graph

✔ energy can be transferred to kinetic energy during an explosion.

You should be able to:

✔ define and understand the meaning of momentum

✔ apply the law of conservation of momentum to analyse collisions and explosions for motion along a straight line

✔ identify collisions as elastic or inelastic.

Momentum is a conserved quantity. The momentum of a system is never lost or gained in a collision unless external forces act on it.

> **momentum** = *mass × velocity*
>
> Velocity is a vector quantity; mass is a scalar, so momentum is also a vector – always specify both its magnitude and direction.
>
> The fundamental unit of momentum is $kg\,m\,s^{-1}$; this is equivalent to the newton second ($N\,s$).

The *law of conservation of momentum* states that, when no external forces act on a system, the vector sum of the momenta before the collision is equal to the vector sum after the collision:

$$m_1 \times u_1 + m_2 \times u_2 + \ldots = m_1 \times v_1 + m_2 \times v_2 + \ldots$$

where m_1, m_2 … are the masses of the objects u_1, u_2 … are the initial velocities and v_1, v_2… are the final velocities of the objects.

This can also be expressed as $\sum mv = 0$, meaning that the sum of the *mass × velocity* for each object before and after the collision must be 0.

A girl throws a ball high into the air. She remains stationary but the force she exerts on the ball causes it to accelerate and gain vertical speed. The ball gains momentum because an external force has acted on it.

When the system consists of the Earth plus the girl and ball, the ball still gains upward momentum. However, the girl also increases her downwards force on the ground as she throws giving momentum to the Earth, which recoils in the opposite direction to the ball's velocity. These momenta are equal and opposite.

Example 2.4.1

An apple is released from rest and falls towards the surface of Earth. Discuss how the conservation of momentum applies to the Earth–apple system.

Solution
The points to make are:

- the forces on the Earth and the apple are equal and opposite

- no external force acts on this isolated system

- changes in the momentum of the Earth and the apple are equal and opposite

- the momentum of the Earth–apple system stays the same and is conserved.

> **>> Assessment tip**
>
> There are two parts to this law.
>
> - There is no change in momentum during a collision.
>
> - Providing no **external** forces act, internal forces do not make any difference as they must act equally (N3) on all parts of the object.
>
> Quote both parts when writing about this law.

> **Newton's second law of motion (N2)** can be written in terms of change in velocity:
>
> $$F = m \times \left(\frac{\Delta v}{\Delta t} \right), \text{ where } \Delta \text{ means}$$
>
> "change in".
>
> This can be rewritten (provided mass is constant) as
>
> $$F = \frac{(m \times \Delta v)}{\Delta t} = \frac{\Delta (\text{momentum})}{\Delta t}$$
>
> which means
>
> $$\frac{\text{change in momentum}}{\text{change in time}}.$$
>
> This interpretation of N2 shows that force acting is the rate of change of momentum.

Kinetic energy E_k is

$$\frac{1}{2}mv^2 = \frac{1}{2} \times \frac{1}{m}(mv)^2 = \frac{p^2}{2m},$$

where p is the momentum of a mass m.

Example 2.4.2

An air-rifle pellet is fired into a wooden block resting on a rough table.

The pellet and the block then slide along the table in a straight line for a distance of 2.8 m before coming to rest.

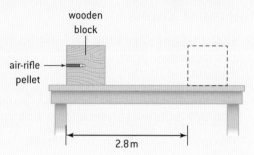

The speed of the block immediately after the collision is $4.8\,\mathrm{m\,s^{-1}}$.

The mass of the pellet is 2.0 g.

The mass of the block is 56 g.

a) Determine the speed of impact of the pellet.

b) Compare the initial kinetic energy of the pellet with the kinetic energy of the pellet and block immediately after the collision.

Solution

a) The initial momentum of the system (pellet + block) is $0.056 \times 0 + 0.002 \times u_1$. The initial velocity of the pellet is u_1 and movement to the right is taken to be positive.

The final momentum of the system is $0.058 \times 4.8 \ \mathrm{kg\ m\ s^{-1}}$.

The initial and final momenta are equal.

This means that $u_1 = \dfrac{5.8 \times 10^{-2} \times 4.8}{2.0 \times 10^{-3}} = 139\ \mathrm{m\,s^{-1}}\ (= 140\,\mathrm{m\,s^{-1}}$ to 2 significant figures).

b) The initial kinetic energy is $\dfrac{1}{2} \times 2.0 \times 10^{-3} \times 139^2 = 19.3\ \mathrm{J}$.

The final kinetic energy of block and pellet is

$$\frac{1}{2} \times 58 \times 10^{-3} \times 4.8^2 = 0.67\ \mathrm{J}.$$

About 18.6 J (19.3 J – 0.67 J) of the energy is transferred from kinetic energy during the collision. This energy will appear as sound, deformation of the wood of the block and some thermal energy. As the block slows down later, all the kinetic energy will eventually transfer to a thermal form.

Even though momentum is conserved in all collisions, the kinetic energy of the moving objects is reduced because of the collision. Example 2.4.3 illustrates this.

A collision in which kinetic energy is removed from the system is known as *inelastic*. When energy is conserved in a collision, it is said to be *elastic*. In other collisions, such as the release of a compressed spring (Example 2.4.3) or when a stationary gun is fired, kinetic energy appears during the collision.

Example 2.4.3

Two masses, m and M, on a frictionless horizontal table are connected by a compressed spring and released.

Mass m moves with velocity v_m.

Mass M moves with velocity v_M.

a) State the change in the momentum of mass m.

b) Deduce the change in momentum of mass M.

c) Deduce the change in energy of the masses.

Solution

a) The initial momentum of the system (both masses) is zero; the change in momentum of m is $m \times v_m$.

b) initial momentum = final momentum

So $0 = mv_m + Mv_M$. Therefore, $v_M = -\dfrac{mv_m}{M}$

c) The initial energy is zero. The total final energy is:

$$\frac{1}{2}mv_m^2 + \frac{1}{2}Mv_M^2 = \frac{1}{2}\left(mv_m^2 + \frac{Mm^2v_m^2}{M^2}\right) = \frac{1}{2}m\left(\frac{m+M}{M}\right)v_m^2$$

This is also the change in energy.

Impulse links to a graph of the variation with time of the force acting on an object. The graph in Example 2.4.4 is typical of that often seen when one object collides with another.

> **Impulse** is the change in momentum of an object and is equal to *force × time for which the force acts.*
>
> The unit of impulse is N s. In fundamental units, this is $kg\,m\,s^{-1}$.
>
> The **area under a graph of force against time** is equal to the impulse and can be used to estimate the momentum change when a force acts on an object.

Example 2.4.4

A ball of mass 0.075 kg strikes a vertical wall with a horizontal velocity of 2.2 m s^{-1} and rebounds horizontally. The time for the impact is 90 ms.

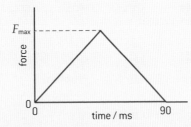

During the collision, 25% of the ball's initial kinetic energy is transferred to other energy stores.

a) Determine the rebound speed of the ball from the wall.

b) Determine the impulse given to the ball by the wall.

c) The graph shows how the force F exerted by the wall on the ball varies with time.

Estimate F_{max}.

Solution

a) The initial kinetic energy of the ball is $\dfrac{1}{2} \times 0.075 \times 2.2^2 = 0.182$ J.

25% of this energy is transferred from kinetic energy.

So, $\dfrac{0.182 \times 3}{4} = 0.136$ J remains.

This gives a rebound speed v of $\sqrt{\dfrac{2 \times 0.136}{0.075}} = 1.90$ m s^{-1}.
The direction is to the left.

b) The impulse given to the wall by the ball is equal to the change in momentum of the ball and is in the same direction. This is $m(v - u)$. However, care is needed with the sign as u is the initial speed to the right and v is the final speed to the left. Therefore, the change of momentum *to the left* is
$0.075 \times (1.9 - (-2.2)) \equiv 0.075 \times 4.1 = 0.31\,\mathrm{N\,s}$ to the left.

c) The impulse equals the area under the force–time graph.
$$0.31 = \frac{1}{2} \times 0.090 \times F_{max} \text{ so } F_{max} = 6.8\,\mathrm{N}.$$

Practice problems for Topic 2

Problem 1
An object is thrown vertically upwards at the edge of a vertical sea cliff. The initial vertical speed of the object is $16\,\mathrm{m\,s^{-1}}$ and it is released 95 m above the surface of the sea.

Air resistance is negligible.

a) Calculate the maximum height reached by the object above the sea.

b) Determine the time taken for the object to reach the surface of the sea.

Problem 2
An object is at rest at time $t = 0$. The object then accelerates for 12.0 s at $1.25\,\mathrm{m\,s^{-2}}$.

Determine, for time $t = 12\,\mathrm{s}$:

a) the speed of the object

b) the distance travelled by the object from its rest position.

Problem 3
An automobile of mass 950 kg accelerates uniformly from rest to $33\,\mathrm{m\,s^{-1}}$ in 11 s.

a) Calculate the resultant force exerted by the automobile to produce this acceleration.

b) The manufacturer claims a maximum speed of 180 km per hour for the automobile. Explain why an automobile has a maximum speed.

Problem 4
A bus travels at constant speed of $6.2\,\mathrm{m\,s^{-1}}$ along a road inclined upwards at $6.0°$ to the horizontal. The mass of the bus is $8.5 \times 10^3\,\mathrm{kg}$. The total output power of the engine of the bus is 70 kW and the efficiency of the engine is 35%.

a) Draw a labelled sketch to represent the forces acting on the bus.

b) Calculate the input power to the engine.

c) Determine the rate of increase of gravitational potential energy of the bus.

d) Estimate the magnitude of the resistive forces acting on the bus.

e) The engine of the bus stops working.

(i) Determine the magnitude of the net force opposing the motion of the bus at the instant at which the engine stops.

(ii) Discuss, with reference to the air resistance, the change in the net force as the bus slows down.

Problem 5
A hammer drives a nail into a block of wood.

The mass of the hammer is 0.75 kg and its velocity just before it hits the nail is $15.0\,\mathrm{m\,s^{-1}}$ vertically downwards. After hitting the nail, the hammer remains in contact with it for 0.10 s. After this time, both the hammer and the nail have stopped moving.

a) Deduce the change in momentum of the hammer during the time it is in contact with the nail.

b) Calculate the force applied by the hammer to the nail.

Problem 6
A magazine article suggests that wearing seat belts in vehicles can save lives in collisions.

Explain, using the concept of momentum, why this is correct.

3 THERMAL PHYSICS

3.1 TEMPERATURE AND ENERGY CHANGES

You must know:

✔ the molecular theory of solids, liquids and gases

✔ the meaning of internal energy and its connection to the concepts of temperature and absolute temperature

✔ what is meant by a phase change

✔ what is meant by a temperature scale

✔ that specific heat capacity is the energy required to change the temperature of 1 kg of a substance by 1 K when there is no change of state and that specific latent heat is the energy required to change the state of 1 kg of a substance

✔ how to calculate energy differences involving specific latent heat changes and specific heat capacity changes and apply these concepts experimentally.

You should be able to:

✔ describe temperature change in terms of internal energy

✔ describe a phase change in terms of molecular behaviour

✔ describe the molecular differences between solids, liquids and gases

✔ use Kelvin and Celsius temperature scales and convert between them

✔ sketch and interpret graphs showing how the temperature of a substance varies with time and with energy transferred to or from the substance.

When energy is transferred to a substance, its *temperature* rises. When the substance cools, energy is transferred away from it. Temperature is sometimes described as the 'degree of hotness' of a body. At the microscopic level, temperature is related to the motion of the atoms and molecules. Temperature can be identified as the mean kinetic energy of particle in the ensemble.

Temperature is defined using temperature scales. The present-day scientific scale is the *Kelvin scale*. It is also known as the absolute temperature scale. A temperature scale requires two *fixed points* (temperatures defined in terms of the properties of a substance).

The connection between temperature and the motion of the particles in a substance gives a direct link between the macroscopic and microscopic energy descriptions.

When the average kinetic energy of one group of particles is higher than that of a second group, the first group has a higher temperature. In mathematical terms, the kinetic energy of the average particle is given by $E_k = \frac{3}{2}kT$, where k is the Boltzmann constant and T is the absolute (kelvin) temperature.

Physicists use macroscopic and microscopic models. In Topic 3.1, the behaviour of bulk materials can be modelled using latent heat and heat capacity. Alternatively, the kinetic theory of Topic 3.2 models the motion of atoms and molecules—imagined as infinitesimally small particles.

Use the differences and similarities of such models to assist your learning.

>> **Assessment tip**

Celsius temperature is always expressed in °C. The kelvin unit never has a ° sign and is written K. Temperature differences are either written as K or deg (meaning 'change of degrees'). When you are giving an answer that is a change in temperature, never use °C.

Temperature is a measure of the average kinetic energy of a collection of moving atoms and molecules.

For the **Kelvin scale**, the two **fixed points** are 0 K and 273.16 K. These are **absolute zero** (the temperature at which atoms and molecules have no kinetic energy, equal to −273.15 °C), and the **triple point of water** (equal to 0.16 °C). The triple point is where all three phases of water co-exist in a sealed container; it occurs at a unique temperature and pressure.

Kelvin defined the degree to be identical in both kelvin and Celsius. To convert from °C to K, add 273. To convert from K to °C, subtract 273.

The **internal energy** of a substance is equal to the sum of the random **kinetic energy** of a collection of particles plus the **potential energy** that arises from intermolecular forces.

The concept of an ideal gas is covered in Topic 3.2.

The term **particle** describes atoms and molecules and models them as small point-like objects without size or shape.

The potential-energy contribution to *internal energy* appears only in liquids and solids where the substances are bound by intermolecular forces. Gases with two or more atoms per molecule can have rotational and vibrational energies as well.

Example 3.1.1

Outline the difference in internal energy of a piece of metal and the internal energy of an ideal gas.

Solution
The internal energy of the metal equals the total kinetic energy of the atoms plus the potential energy of the system. This potential energy arises from the bonds between the metal atoms.

The molecules of an ideal gas have only kinetic energy. One of the assumptions of an ideal gas is that the particles in it do not interact through molecular bonds and so there is no potential energy to consider.

Example 3.1.2

Sketch a graph to show the relationship between the internal energy of an ideal gas and its temperature measured in degrees Celsius. Explain the key features of this graph.

Solution

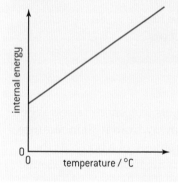

The graph indicates that, at 0 °C, there is some internal energy in the gas. This is because the internal energy of the gas is directly proportional to the absolute temperature. Absolute zero is at 0 K and this is the intercept on the temperature axis; in Celsius, this value is −273 °C.

There are four known *states of matter*. One of these, plasma, only exists at high temperatures. The other three states are *solid*, *liquid* and *gas*.

All substances are made up of atoms and molecules and the differences between states is due to the nature of the bonding between them.

Normally, when thermal energy is transferred to a solid, it first becomes a liquid and then changes to a gas. These are called *phase changes*.

The phase change names link to the latent heat changes later in this topic. Phase changes used in the DP Physics course are:

- melting and freezing (changes between solid and liquid)

- boiling and condensing (changes between liquid and gas).

Phase changes can be demonstrated by transferring thermal energy to a solid. The graph in Figure 3.1.1 shows the variation of temperature with energy transferred.

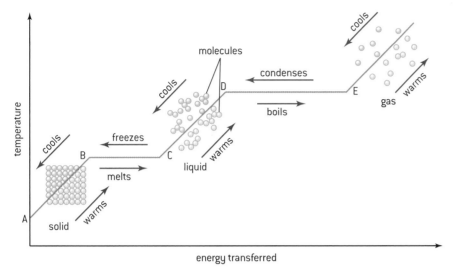

▲ **Figure 3.1.1.** Variation of temperature with energy transferred

The graph can also be plotted as *temperature* against *time* (when the energy is input at a constant rate).

When the state of a substance does not change, its temperature rises—a *heat capacity* change. The internal energy of the substance is increasing while potential energy is largely unchanged (this is only approximately true, especially when expansion or contraction occurs).

While the state changes, the temperature is constant. Energy is being transferred into the potential form and the kinetic energy is constant—a *latent heat* change.

Example 3.1.3

The internal energy of a piece of zinc is increased by 1.5 kJ by heating. Its subsequent increase in temperature is 11 deg. The piece of zinc has mass 0.35 kg.

a) Explain the meaning of *internal energy* and *heating*.

b) Calculate the specific heat capacity of zinc.

Solution

a) The internal energy is the sum of the potential energy and the kinetic energy of the zinc atoms. It can also be described as the amount of energy stored in the zinc.

Heating is the process of transferring energy using a non-mechanical or thermal pathway from an energy source to the zinc. The zinc is acting as an energy sink.

b) $c = \dfrac{Q}{m\Delta T} = \dfrac{1500}{0.35 \times 11} = 390\,\text{J kg}^{-1}\text{K}^{-1}$

Gases have individual particles that are independent of each other and move freely within a container, filling it completely. Pressure arises as the particles interact with the container walls.

Liquids can move around within the bulk of the material. The particles only interchange with nearest neighbours, which enables a liquid to have a definite volume but be able to flow.

Solids allow little, if any, movement between particles, which rarely exchange positions with each other. Solids have a fixed shape.

The **specific heat capacity** c of a substance is the energy required to change 1 kg of the substance by 1 deg (or 1 K). The unit of specific heat capacity is J kg^{-1} K^{-1}. In fundamental units, this is m^2 s^{-2} K^{-1}.

$$c = \frac{Q}{m \times \Delta T}$$

where

Q is the energy transferred (in J)

m is the mass (in kg)

ΔT is the change in temperature (in K)

The **specific latent heat** L of a substance is the energy required to change the phase of 1 kg of the substance. The unit of specific latent heat is J kg^{-1}. In fundamental units, this is m^2 s^{-2}.

$$L = \frac{Q}{m}$$

To specify a specific latent heat change, state the type of phase change that is occurring. For example, "The specific latent heat of freezing of ice (in other words, going from water to ice) is 0.34 MJ kg^{-1}".

The specific heat capacity of a material can be determined using the *method of mixtures*. A hot solid of known temperature is added to a cold liquid, also at known temperature. The masses of the solid and liquid are known. The resulting mixture of solid and liquid reaches a final, measured temperature. When one of the specific heat capacities is known, the other can be determined.

Present any multi-step solution clearly with a clear description of each step. An examiner can then give you partial credit if you have made an error elsewhere.

SAMPLE STUDENT ANSWER

In an experiment to determine the specific latent heat of fusion of ice, an ice cube is dropped into water contained in a well-insulated calorimeter of negligible specific heat capacity. The following data is available.

Mass of ice cube	$= 25\,g$
Mass of water	$= 350\,g$
Initial temperature of ice cube	$= 0\,°C$
Initial temperature of water	$= 18\,°C$
Final temperature of water	$= 12\,°C$
Specific heat capacity of water	$= 4200\,J\,kg^{-1}\,K^{-1}$

a) Using the data, estimate the specific latent heat of fusion ice.   [4]

This answer could have achieved 2/4 marks:

▲ There are correct calculations of the energy lost by the cooling water and the (heat capacity) energy gained by the ice once it has melted. Then there is a recognition that 7.56 kJ is the energy available to melt the ice.

$$Q = mc\Delta T \qquad m = 0.350\,kg \qquad c = 4200\,J/kg.K \qquad \Delta T = 6\,°k$$

$$Q = 0.350\,kg \times 4200 \times 6\,°k = 8.82\,kJ$$

$$m = 0.025\,kg \qquad \Delta = 12\,°C$$

$$0.025\,kg \times 4200 \times 12\,°k = 1.26\,kJ \qquad c = 4200\,J/kg.K$$

$$7.56\,kJ = 0.375 \times 4200 \times ?$$

$$\frac{7.56\,kJ}{1.575\,kg.K} = 4.8\,°K + 273 = 277.8\,°K$$

▼ The equation is incorrect (it is for heat capacity not latent heat).

b) The experiment is repeated using the same mass of ice. This time, the ice is crushed.

Suggest the effect of this, if any, on the time it takes the water to reach its final temperature.   [1]

This answer could have achieved 0/1 marks:

The time it takes to reach the final temperature will decrease.

▼ The answer is correct, but the answer fails to appreciate the significance of the command term 'suggest'. This means to propose a hypothesis and it requires some explanation of the proposal. The answer needs to go on to say that the surface area of the ice increases when crushed, so the water and the ice can interact more quickly.

3.2   MODELLING A GAS

You must know:

✔ what is meant by pressure, ideal gas, mole, molar mass and the Avogadro constant

✔ the equation of state for an ideal gas

✔ the gas laws and an experimental investigation of one gas law

✔ the differences between an ideal and a real gas

✔ that gas laws are empirical and that the kinetic model is theoretical.

You should be able to:

✔ sketch and interpret p–V, P–T and V–T graphs to interpret changes of state of a gas

✔ solve problems using the equation of state for an ideal gas

✔ investigate at least one gas law experimentally

✔ understand aspects of the molecular kinetic model of an ideal gas: the assumptions of the model and how they lead to a theoretical model.

Pressure arises with all three phases of matter. The pressure on the walls of a gas container depends on the rate at which the gas particles transfer momentum when they collide with the wall.

Quantity of matter is the way scientists compare numbers of objects. A *mole* of atoms, a mole of electrons, a mole of ions, always gets you about 6.022×10^{23} objects (atoms, electrons and ions respectively).

A mole of atoms of a chemical element has a mass equal to its atomic mass number in grams. For example, one mole of the isotope of carbon-12 $\left({}^{12}_{6}C\right)$ as a mass of 12 g. This is known as the *molar mass*.

The behaviour of gases at extremes of temperature and pressure is complicated. A simplified model of an *ideal gas* is used in which the particles collide elastically and in which no intermolecular forces act. For a *real gas*, there are effects between molecules and with the walls. Within a few degrees of absolute zero, other effects become important to make the gas non-ideal.

Pressure is defined as $\dfrac{\text{force}}{\text{area}}$. Its unit is $N\,m^{-2}$, which is the same as a pascal, Pa. In fundamental units, this is $kg\,m^{-1}\,s^{-2}$.

In solids, a normal force applies through a contact area between the solid and the surface on which it rests.

Real gases have behaviour close to ideal only for low pressures, low densities and moderate temperatures. Treat gases as ideal unless told otherwise.

An **equation of state** describes a gas using three variables: pressure p, volume V and temperature T. Equations of state are possible for real gases but need extra terms to account for high densities and particle interactions.

The **equation of state for an ideal gas** is

$pV = nRT$

where n is the number of moles.

This can be summarized in the general gas equation for two states, 1 and 2, of a gas:

$$\frac{p_1V_1}{T_1} = \frac{p_2V_2}{T_2}$$

The equation can also be written as

$$pV = \frac{N}{N_A}RT \quad \text{and} \quad pV = Nk_BT$$

where

R is the (ideal) **gas constant** ($8.31\,J\,mol^{-1}\,K^{-1}$)

k_B is the **Boltzmann constant** ($1.38 \times 10^{-23}\,J\,K^{-1}$)

N is number of molecules.

There is more detail on pressure in a liquid in Option B.3.

The **mole** is the fundamental SI unit for the quantity of matter of a substance. It corresponds to the mass of a substance that contains 6.022×10^{23} particles of the substance.

The number 6.022×10^{23} is known as the **Avogadro constant** N_A.

The number of moles n of a substance $=$ $\dfrac{\text{number of molecules}}{N_A} = \dfrac{N}{N_A}$.

Example 3.2.1

An ideal gas in a container of volume $1.2 \times 10^{-5}\,m^3$ has a pressure of $1.5 \times 10^5\,Pa$ at a temperature of 50 °C.

Calculate the number of molecules of gas in the container.

Solution
The temperature must be in kelvin: $50 + 273 = 323\,K$.

$$n\left(= \frac{pV}{RT}\right) = \frac{1.5 \times 10^5 \times 1.2 \times 10^{-5}}{8.31 \times 323} = 6.71 \times 10^{-4}\ \text{mol}$$

The solution requires the number of molecules $= nN_A$

$$= 6.71 \times 10^{-4} \times 6.02 \times 10^{23} = 4.04 \times 10^{20}\ \text{molecules}$$

>> Assessment tip

Notice that both sides of the equation of state have the units of energy. You can think of R as being analogous to the specific heat capacity of one mole of a gas, with k_B being analogous to the specific heat capacity of one molecule.

You must know the details of one experimental investigation of a gas law. This might be tested in Paper 1, 2 or 3.

Historically, gas behaviour was identified through experiments involving three separate *gas laws* that, taken together, reflect the *equation of state*.

>> Assessment tip

When using the general gas equation, the units of pressure and volume must match on both sides of the equation. The **only** unit allowed for temperature in the equation is kelvin.

These graphs summarize the essential features of the gas laws.

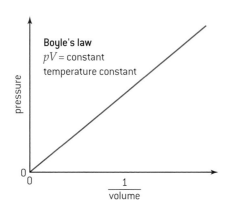

Boyle's law
pV = constant
temperature constant

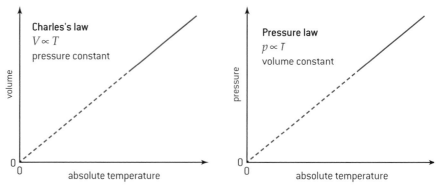

Charles's law
$V \propto T$
pressure constant

Pressure law
$p \propto T$
volume constant

The gas laws were suggested by scientists in the 18th century following experimental work. Such laws are said to be **empirical**. On the other hand, the kinetic model of a gas stems from a **theoretical** standpoint involving assumptions about the gas. These macroscopic and microscopic approaches fit together confirming our view of gas properties.

The kinetic model of a gas is based on the following assumptions about gas particles and their behaviour.

- A gas consists of particles; the total volume of the particles is negligible compared with the total volume of the gas (or you could say, the average distance between particles is greater than their individual size).

- Particles have the same mass.

- Particles are in constant, random motion.

- Particles collide elastically with each other and the walls of the container.

- Interactions between particles can be ignored (so they do not exert force on each other).

- The time for a particle collision is negligible compared with the time between collisions.

- Gravity can be ignored.

>> **Assessment tip**

Know the meaning of these assumptions and recognise how they affect the kinetic model. The model leads to the relationship between the pressure of the gas and the mean square speed of the particles. The steps in the derivation of the model are given in Example 3.2.2.

Example 3.2.2

A particle of mass m moves with velocity u in a box with side lengths x, y and z.

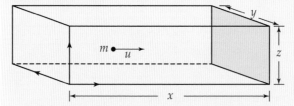

The particle strikes the end faces at right angles and makes repeated elastic collisions with opposite faces.

a) i) Calculate the time t between collisions with the shaded face.

ii) State the expression for the change in momentum per collision with the shaded face.

b) The box contains N identical particles, all moving parallel to the x-direction with speed u_x, making elastic collisions at the ends. Determine the average force F on the shaded face.

c) The model is refined so that N particles of average squared speed $\overline{c^2}$ move randomly in the box.

The speed in direction y is v.

The speed in direction z is w.

i) Deduce an expression for u^2 in terms of $\overline{c^2}$.

ii) Deduce an expression for F in terms of $\overline{c^2}$.

iii) Show that $pV = \dfrac{Nm\overline{c^2}}{3}$, where p is the pressure of the gas and V is its volume.

Solution

a) i) The particle travels a distance $2x$ at a speed u, so $t = \dfrac{2x}{u}$.

ii) The change in momentum for each collision at the shaded face $= \Delta p = 2mu$.

b) force = rate of change of momentum $= \left(\dfrac{\Delta p}{\Delta t}\right) = \dfrac{2mu}{\left(\dfrac{2x}{u}\right)}$.

So, for N particles, force $F = N\left(\dfrac{\Delta p}{\Delta t}\right)$, which is:

$N \times 2mu \times \dfrac{u}{2x} = \dfrac{Nmu^2}{x}$.

c) i) For one particle, the magnitude of its velocity c is given by
$c^2 = u^2 + v^2 + w^2$.

For all the particles, $\overline{c^2} = \overline{u^2} + \overline{v^2} + \overline{w^2}$

ii) The average squared speeds must be the same in each direction as we do not see directional differences within gases.

Therefore, $\overline{u^2} = \dfrac{\overline{c^2}}{3}$ and $F = \dfrac{1}{3}\dfrac{Nm\overline{c^2}}{x}$.

iii) $p = \dfrac{F}{A} = \dfrac{1}{3}\dfrac{Nm\overline{c^2}}{x} \times \dfrac{1}{yz} = \dfrac{1}{3}\dfrac{Nm\overline{c^2}}{xyz}$ but xyz = volume of box V

Therefore, $p = \dfrac{1}{3}\dfrac{Nm\overline{c^2}}{V}$ or $pV = \dfrac{Nm\overline{c^2}}{3}$

The equation of state links the pressure and volume of a gas to the speeds of the particles. The temperature can also link directly to the average squared speeds:

$Nk_BT = \dfrac{Nm\overline{c^2}}{3}, T = \dfrac{m\overline{c^2}}{3k_B}$ and therefore $\dfrac{3}{2}k_BT = \dfrac{1}{2}m\overline{c^2}$.

The right-hand side is the average kinetic energy of a gas particle. The left-hand side is a measure of the kinetic energy of a gas molecule $\left(\overline{E_K}\right)$. The units of k_BT are equivalent to joules.

Example 3.2.3

A cylinder of fixed volume contains 15 mol of an ideal gas at a pressure of 490 kPa and a temperature of 27 °C.

a) Determine the volume of the cylinder.

b) Calculate the average kinetic energy of a gas molecule in the cylinder.

Solution

a) Use $pV = nRT$ to give $V = \dfrac{15 \times 8.31 \times (27 + 273)}{4.9 \times 10^5} = 0.076\,\text{m}^3$.

b) Use $\overline{E_K} = \dfrac{3}{2}k_BT$. $\overline{E_K} = \dfrac{3}{2} \times 1.38 \times 10^{-23} \times 300 = 6.2 \times 10^{-21}\,\text{J}$

SAMPLE STUDENT ANSWER

0.46 mol of an ideal monatomic gas is trapped in a cylinder. The gas has a volume of 21 m³ and a pressure of 1.4 Pa.

a) State how the internal energy of an ideal gas differs from that of a real gas. [1]

This answer could have achieved 0/1 marks:

Internal energy is constant as molecules move at constant velocity.

▼ The kinetic model does not assume constant velocity and this answer does not make it clear whether it is referring to the ideal or the real case.

This answer could have achieved 1/1 marks:

Ideal gas ignores intermolecular force between molecules in between collision. So there is no potential energy, and contains kinetic energy only.

▲ The distinction between ideal and real gases is clear even though the word 'real' does not appear in the answer. The deduction that no intermolecular force implies no potential energy is correct and is reinforced by the statement about kinetic energy.

b) Determine, in kelvin, the temperature of the gas in the cylinder. [2]

This answer could have achieved 2/2 marks:

$pV = nRT$

$1.4 \times 21 = 0.46 \times 8.31 \times T$

$T = \dfrac{29.4}{3.8226}$

$= 7.7$ Kelvin

The temperature of the gas is 7.7 Kelvin.

▲ The answer begins with a clear statement of the gas equation to be used. The substitution is clear (because the numbers are substituted in the same order as in the equation) and the answer is quoted to an appropriate number of significant figures.

Practice problems for Topic 3

Problem 1

a) X and Y are two solids with the same mass at the same initial temperature. Their temperatures are raised by the same amount; they both remain solid.

The specific heat capacity of X is greater than that of Y.

Explain which substance has the greater increase in internal energy.

b) Cold water, initially at a temperature of 14 °C, flows over an insulated heating element in a domestic water heater. The heating element transfers energy at a rate of 7.2 kW. The water leaves the heater at a temperature of 40 °C.

The specific heat capacity of water is $4.2 \, kJ \, kg^{-1} \, K^{-1}$.

(i) Estimate the rate of flow of the water.

(ii) Suggest **one** reason why your answer to part b) i) is an estimate.

Problem 2

a) Distinguish between thermal energy and internal energy.

b) Outline, with reference to the particles, the difference in internal energies of a metal and an ideal gas.

Problem 3

Use the kinetic model to explain why:

a) the pressure of an ideal gas increases when heated at constant volume

b) the volume of an ideal gas increases when heated at constant pressure.

Problem 4

A quantity of 0.25 mol of an ideal gas has a pressure of $1.05 \times 10^5 \, Pa$ at a temperature of 27 °C.

a) Calculate the volume occupied by the gas.

b) When the gas is compressed to $\dfrac{1}{20}$ of its original volume, the pressure rises to $7.0 \times 10^6 \, Pa$.

Calculate the temperature of the gas after the compression.

Problem 5

The pressure in a container is increased using a bicycle pump. The volume of the container is $1.30 \times 10^{-3} \, m^3$.

The pump contains $1.80 \times 10^{-4} \, m^3$ of air at a pressure of 100 kPa and a temperature of 300 K.

Assume that the air acts as an ideal gas.

Assume that all the air molecules from the pump are transferred into the container when the pump is pushed in.

a) The air in the container is at an initial pressure of 150 kPa and a temperature of 300 K.

(i) Calculate, in mol, the initial quantity of gas in the container.

(ii) Calculate, in mol, the quantity of gas transferred to the container every time air is pumped into it.

(iii) The temperature of the gas in the pump returns to 300 K after the pump has been used.

Calculate the pressure in the container after the pump has transferred one pump-full of air into the container and its temperature has returned to 300 K.

b) Explain, with reference to the kinetic model of an ideal gas, why the gas in the container has pressure and why this pressure will increase when gas molecules are transferred to the container.

Problem 6

Air in a container has a density of $1.24 \, kg \, m^{-3}$ at a pressure of $1.01 \times 10^5 \, Pa$ and a temperature of 300 K.

a) Calculate the mean kinetic energy of an air molecule in the container.

b) Calculate the mean square speed for the air molecules.

c) The temperature of the air in the container is increased to 320 K.

Explain why some of the molecules will have speeds much less than that calculated in part (b).

4 OSCILLATIONS AND WAVES

4.1 OSCILLATIONS

You must know:

✔ what is meant by an oscillation

✔ the definitions of time period, frequency, amplitude, displacement, phase difference

✔ the conditions for simple harmonic motion.

✔ the relationship between acceleration and displacement in simple harmonic motion.

You should be able to:

✔ sketch and interpret graphs for simple harmonic motion of displacement–time, velocity–time, acceleration–time and acceleration–displacement

✔ describe the energy changes that take place in one cycle of an oscillation.

Time period T is the time for one cycle of the oscillation. Its unit is the second (s).

Frequency f is the number of cycles of the oscillation in one second. Its unit is the hertz (Hz or s^{-1}).

Time period and frequency are connected by $f = \dfrac{1}{T}$.

Amplitude x_0 is the maximum displacement of the oscillating object from its equilibrium position. This can be expressed as an angle or a distance.

Displacement x is the distance between the equilibrium and instantaneous positions and, as a vector requires a direction, can be positive or negative.

The **equilibrium position** is the position to which the system returns when it is not oscillating.

Phase difference is the difference, in degrees or radians, between two oscillations at the same instant in time.

🔗 There is more about the mathematics of shm in HL Topic 9.1. Concepts in Topic 4 are directly connected to the mechanics of Topic 2.

A pendulum—a mass swinging at the end of a string—is an example of an oscillating system. A *cycle* for this system is the movement of the mass from the rest position at one end of the swing, through to the opposite side and back to the original rest position. Motion from one side to the other is half a cycle.

The rest position (in the middle of the swing for a pendulum) is also known as the equilibrium position which is the position where the system will be when not oscillating.

Simple harmonic motion (shm) is an oscillation for which

acceleration ∝ −displacement

For **simple harmonic motion**:

• the acceleration of the object is directly proportional to its displacement

• the vector direction of acceleration is opposite to the displacement; this is the meaning of the negative sign in the equation.

Example 4.1.1

A mass, with an equilibrium position at O, is displaced to point X and released from rest. Its motion is simple harmonic.

Identify where the acceleration of the mass is greatest.

Solution
Because **acceleration** is proportional to **−displacement**, the greater the distance from O, the greater the magnitude of the acceleration. So, acceleration is greatest at X.

Graphs for shm, showing the variation of acceleration, velocity and displacement with time, are given in Figure 4.1.1.

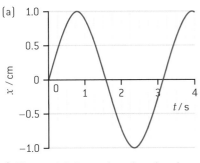

(a)

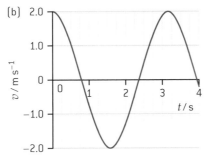

(b)

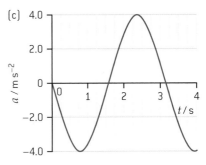

(c)

▲ Figure 4.1.1. x–t, v–t and a–t graphs for simple harmonic motion

Instantaneous velocity at a particular time is equal to the gradient of the displacement–time graph.

Instantaneous acceleration is equal to the gradient of the velocity–time graph.

Mathematically, $v = \dfrac{\Delta s}{\Delta t}$ and $a = \dfrac{\Delta v}{\Delta t}$ (Δ means 'change in').

This makes the a–t graph the inversion of the x–t graph.

This is expected, since acceleration $\propto$ −displacement (Figure 4.1.2).

• The v–t graph lags the x–t graph by 90° (or is 90° out of phase).

• The a–t graph is 180° out of phase with the x–t graph.

Energy transfers occur throughout the oscillator cycle. For a mass–spring system, the kinetic energy of the mass transfers to and from elastic potential energy in the spring. This energy transfer sustains the oscillation indefinitely when no friction acts.

Figure 4.1.3 shows the energy variations plotted with time.

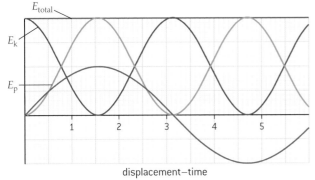

displacement–time

▲ Figure 4.1.3. Energy variations against time for shm

Example 4.1.2

A mass hanging on a spring is pulled vertically down 0.15 cm from the equilibrium position and released. The mass returns to the equilibrium position 0.75 s after release.

State:

a) the amplitude

b) the time period for the oscillation.

Solution

a) The amplitude is the distance from the equilibrium position to the maximum displacement. This is 0.15 cm.

b) The mass has travelled a $\dfrac{1}{4}$ cycle when it reaches the equilibrium position for the first time after release. The time period is $4 \times 0.75\,\text{s} = 3.0\,\text{s}$.

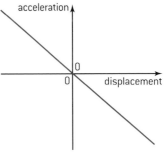

▲ Figure 4.1.2. Acceleration–displacement graph for simple harmonic motion

The energy graphs vary as $\sin^2$ rather than the sine curves of Figure 4.1.1 (a–c). This is covered in more detail in Topic 9.1.

Phase is used extensively in Topic 9.

There are also links to Topic 6 where radian measure and degrees are used: 90° can be written as $\dfrac{\pi}{2}$ rad and 180° written as π rad.

The **kinetic energy** cycle:

• has double the frequency of the motion

• is never negative ($E_k \propto v^2$)

• has a different shape from the sine curves in Figure 4.1.1.

The total energy is constant with time when energy losses are zero. The variation with time of the stored elastic potential energy E_p has the same frequency as the E_k–time graph but is π out of phase.

Example 4.1.3

Which of the following statements is **true** for an object performing simple harmonic motion about an equilibrium position O?

A The acceleration is always away from O.

B The acceleration and velocity are always in opposite directions.

C The acceleration and the displacement from O are always in the same direction.

D The graph of acceleration against displacement is a straight line.

Solution

The correct answer is D.

The acceleration can be in the same or the opposite direction to velocity, so B is incorrect. The negative sign in the definition of shm means that acceleration and displacement are always opposed so A and C cannot be correct. Response D is the alternative way to express this.

>> **Assessment tip**

Multiple-choice questions demand care. The incorrect responses in Example 4.1.3 test the relationships between acceleration, velocity and displacement. The negative sign in $a \propto -x$ means that, because the acceleration and velocity are 90° out of phase, the acceleration and velocity can be in the same direction or opposite.

SAMPLE STUDENT ANSWER

A mass oscillates horizontally at the end of a horizontal spring. The mass moves through a total distance of 8.0 cm from one end of the oscillation to the other.

a) State the amplitude of the oscillation. [1]

This answer could have achieved 0/1 marks:

● 8.0 cm

▼ The student has not visualized the arrangement. The mass is moving from one extreme to the other and passes through the equilibrium position half way through this distance. The amplitude is 4.0cm.

b) Outline the conditions that the system must obey for the motion to be simple harmonic. [2]

This answer could have achieved 2/2 marks:

a must be in the opposite direction to x and a must be proportional to x

▲ These are the two conditions that are required. However it would be better to define the symbols x and a for clarity.

4.2   TRAVELLING WAVES

You must know:

✔ what is meant by wavelength, frequency and wave speed

✔ energy is transmitted in a wave without overall disturbance of the medium

✔ the distinction between transverse and longitudinal waves

✔ the nature of electromagnetic waves

✔ the nature of sound waves

✔ the order of magnitude of wavelengths for the principal regions of the electromagnetic spectrum.

You should be able to:

✔ explain particle motion in a medium that leads to transverse and longitudinal waves

✔ sketch and interpret distance–time and distance–displacement graphs for transverse and longitudinal waves

✔ solve problems involving wave speed, frequency and wavelength

✔ describe an experimental method for investigating the speed of sound.

Waves transfer energy without any overall change in the medium through which they pass.

For *transverse waves*, particles of the medium oscillate at 90° to the direction of energy propagation.

For *longitudinal waves*, particles oscillate in the same direction as the direction of energy propagation.

Longitudinal waves cannot be polarized, but otherwise the physics of both wave types is similar.

Polarization of transverse waves is discussed in Topic 4.3

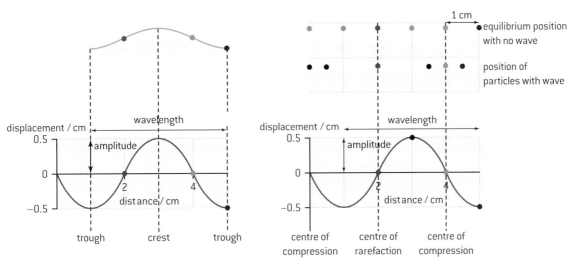

▲ Figure 4.2.1. Displacement–distance graph for transverse and longitudinal waves

A graph of displacement–distance provides a 'snapshot' of the shape of a wave at one moment in time. Figure 4.2.1 shows the graph together with its interpretation for each type of wave.

A representation of the motion of an individual point on the wave is given by a displacement–time graph (Figure 4.2.2). This may look the same as the graph in Figure 4.2.1, but this graph gives a direct value for the time period of the wave, rather than its wavelength.

Wavelength λ is the distance between the two nearest points on the wave with the same phase.

Wave speed c is the speed at which the wave moves in the medium.

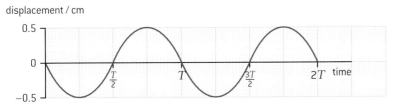

▲ Figure 4.2.2. Displacement–time graph for wave motion

>> **Assessment tip**

Ensure that you read graph axes carefully both for the quantity and the unit.

You need to be able to derive the equation for the speed of a wave:
$c = f\lambda$

Remember that a displacement–distance graph shows that the wave moves forward by λ in one cycle,

whereas a displacement–time graph gives the time T that one particle takes to go through one cycle.

The wave speed is therefore

$\dfrac{\lambda}{T} = f\lambda$ because $f = \dfrac{1}{T}$.

Example 4.2.1

The shortest distance between two points on a progressive transverse wave which have a phase difference of $\frac{\pi}{3}$ rad is 0.050 m. The frequency of the wave is 500 Hz.

Determine the speed of the wave.

Solution

$\frac{\pi}{3}$ rad is 60° which is $\frac{1}{6}$ of a cycle.

This means that the wavelength of the wave is $6 \times 0.050 = 0.30$ m.

The speed of the wave is $c = f\lambda = 500 \times 0.30 = 150 \text{ m s}^{-1}$.

Sound is transmitted through gases and liquids as longitudinal waves. The molecules of a gas are the medium and move as the wave passes through them.

- There are areas of low pressure (rarefactions: below atmospheric pressure) and high pressure (compressions: above atmospheric pressure)

- The positions of maximum and minimum pressure are at the points where the displacement is zero

- The positions of maximum and minimum displacement are where the pressure is atmospheric (in other words, average).

Sound waves show almost all the common properties of waves: reflection, refraction, diffraction and interference, but not polarization (see Topic 4.3).

Example 4.2.2

a) Outline the difference between a longitudinal wave and a transverse wave.

b) State an example of:

    i) a transverse wave   ii) a longitudinal wave.

c) Sound with a frequency of 840 Hz travels through steel with a speed of 4.2 km s^{-1}.

    Calculate the wavelength of the sound wave.

Solution

a) In a transverse wave, the vibrations are perpendicular to the direction of energy propagation.

    In a longitudinal wave, the vibrations of the particles are in the same direction of propagation.

b) i) An electromagnetic wave.   ii) A sound wave in a gas.

c) $c = f\lambda$ so $\lambda = \dfrac{c}{f} = \dfrac{4200}{860} = 4.9$ m

Electromagnetic waves:

- do not need a medium and can travel through a vacuum

- travel at the same speed in a vacuum irrespective of frequency, in other words at $3.0 \times 10^8 \text{ m s}^{-1}$

- have decreased speeds in matter; this speed depends on the frequency.

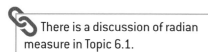
There is a discussion of radian measure in Topic 6.1.

>> **Assessment tip**

Notice that the answer to Example 4.2.2 part b) ii) specifies 'sound wave in a gas'. A solid surface can transmit a sound wave as a transverse wave.

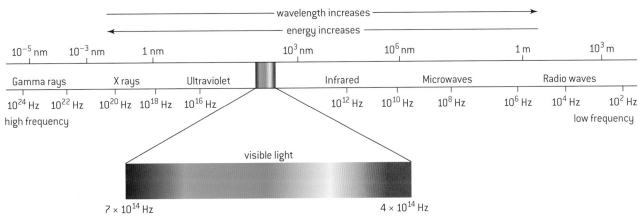

▲ **Figure 4.2.3.** The principal regions of the electromagnetic spectrum and their wavelengths

SAMPLE STUDENT ANSWER

A longitudinal wave is travelling in a medium from left to right. The graph shows the variation with distance x of the displacement y of the particles in the medium. The solid line and the dotted line show the displacement at $t = 0$ and $t = 0.882$ ms, respectively.

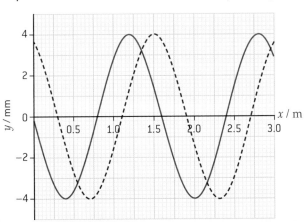

The period of the wave is greater than 0.882 ms. A displacement to the right of the equilibrium position is positive.

a) State what is meant by a longitudinal travelling wave. [1]

This answer could have achieved 1/1 marks:

> A wave where the energy is moving in the same direction as the motion of the particles.

b) Calculate the speed of this wave. [2]

This answer could have achieved 0/2 marks:

> $v = f\lambda \qquad \lambda = 1.6\,(m) \qquad f = \dfrac{\lambda}{T} < 1.13\,Hz$
>
> 0.3 m in 0.882
>
> $V < 1.81\,ms^{-1}$

▶ **Assessment tip**

You should know the order of magnitude of the wavelengths for each region, a use for each wave and a disadvantage of using the wave (which may be a medical disadvantage).

▲ The student has conveyed the sense that this direction is parallel to the displacement of the particles in the wave (although it would be better to talk about the 'direction of propagation of energy').

▼ The graphs show the motion of two particles separated by 0.3 m. The time taken for the maximum amplitude to travel from the particle represented by the solid line to that represented by the dashed line is 0.882 ms.

$\dfrac{\text{displacement}}{\text{time}}$ gives the speed of the wave. There is no credit in this approach which tries to use $\dfrac{1}{0.882}$ as the frequency. Note that there is an additional power of ten error as the time was quoted in milliseconds but used here in seconds.

4.3   WAVE CHARACTERISTICS

You must know:

✔ wave shape is indicated by a wavefront

✔ the meaning of amplitude and intensity

✔ a ray is locally at 90° to a wavefront and shows the direction in which a wave is moving

✔ intensity of a wave is power per unit area

✔ intensity of a wave is proportional to amplitude²

✔ superposition occurs when two or more waves are added together

✔ the meaning of polarization

✔ only transverse waves can be polarized.

You should be able to:

✔ sketch and interpret diagrams involving wavefront and rays

✔ solve problems involving amplitude, intensity and the inverse square law

✔ sketch and interpret the superposition of pulses and waves

✔ sketch and interpret diagrams that illustrate polarized, reflected and transmitted beams

✔ solve problems using Malus's law

✔ calculate the resultant of two waves or pulses using algebra or graphs.

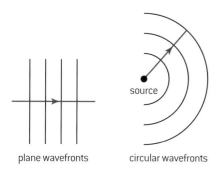

plane wavefronts   circular wavefronts

▲ Figure 4.3.1. Plane and circular wavefronts and rays

A **wavefront** shows the shape of the wave at one instant. **Rays** are at 90° to the wavefront locally and they show the direction of wave movement. A series of successive wavefronts makes it possible to deduce the origin and history of the wave, assuming that wavefronts are one wavelength apart.

Superposition is when two (or more) waves add. This addition is vectorial and you must take account of the *sign* of the displacement. Two waves that have the same displacement but opposite signs give zero displacement overall.

As waves move they can change both their shape and their direction of motion. *Wavefronts* and *rays* are used to visualize these changes (Figure 4.3.1).

When waves travelling in the same medium coincide at the same place in space and time, they *superpose*. Superposition occurs whatever the wavelength or amplitude of the two waves.

The rate of transfer of wave energy is measured using the quantity **intensity**. A point source of a wave spreads out through space. The source radiates power P (this is the energy it transfers in one second). At distance r from the source, this spreads over a sphere of radius r which has an area equal to $4\pi r^2$. The intensity is given by $I = \dfrac{P}{4\pi r^2}$.

The intensity I of a wave is related directly to its amplitude A: $I \propto A^2$.

When a wave has its amplitude doubled, the intensity goes up by 4×, so four times as much energy falls on a given area every second.

Example 4.3.1

A bicycle lamp and a floodlight deliver their output energy into a cone that covers 0.20 of the area of a sphere.

The output power of the bicycle lamp is 1.5 W.

The output power of the floodlight is 300 W.

a) Calculate the intensity of the bicycle lamp when viewed by an observer from 20 m away.

b) The observer has an eye with a pupil diameter of 5.0 mm. Determine the light power entering the eye from the bicycle lamp.

c) The bicycle lamp and the floodlight are equally bright to the observer. Deduce the distance of the floodlight from the observer.

Solution

a) 1.5 W of light is delivered to 0.20 of a sphere of radius 20 m.

$$I = \frac{P}{4\pi r^2} = \frac{1.5}{4 \times \pi \times 20^2 \times 0.2} = 1.5 \times 10^{-3} \, \text{mW m}^{-2}$$

b) The area of the pupil is $\pi \times (2.5 \times 10^{-3})^2 = 1.96 \times 10^{-5} \, \text{m}^2$.

So the total power entering is $1.5 \times 10^{-3} \times 1.96 \times 10^{-5} = 29 \, \text{nW}$.

c) The intensity entering the eye must be the same for both lamps.

$$\frac{P_b}{r_b^2} = \frac{P_f}{r_f^2} \text{ therefore, } \frac{r_b^2}{r_f^2} = \frac{P_b}{P_f} \text{ and } \frac{r_b}{r_f} = \sqrt{\frac{P_b}{P_f}}$$

As the power ratio is $300 : 1.5 = 200 : 1$, the distance ratio is $14 : 1$ and the floodlight is $20 \times 14 = 280 \, \text{m}$ away.

> **» Assessment tip**
>
> Try to use technical language correctly. You may be allowed **superimpose** for **superpose** but if there is doubt about what you mean you will not receive credit.

> **Intensity** is the amount of energy that a wave can transfer to an area of one square metre in one second. The unit of intensity is W m^{-2}. In fundamental units, this is kg s^{-3}.

Polarization is the restriction of the oscillation of the wave to one plane (Figure 4.3.2). Transverse waves can be polarized because the energy propagation is at 90° to the particle displacement; longitudinal cannot because the energy propagation and displacement are parallel.

> When the wavelengths or frequencies superposing are identical, other effects (described in Topics 4.4 and 4.5) are observed.

Polarization has many uses, typically:

- glare reduction in sunglasses

- stress analysis modelling the behaviour of engineering structures

- 3D movies where two films with different polarizations are shown and polarizing lenses are used to isolate the two images.

A polarized wave of intensity I_0 will be reduced in intensity when it is incident on a polarizer with a polarizing axis at θ to that of the original polarizer. Malus's law predicts that the final intensity I of the polarized beam will be $I = I_0 \cos^2 \theta$.

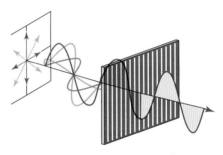

▲ Figure 4.3.2. Polarization of a transverse wave

Example 4.3.2

a) State what is meant by polarized light.

b) Polarized light of intensity I_0 is incident on an analyser. The transmission axis of the analyser makes an angle θ with the direction of the electric field of the light.

Calculate, in terms of I_0, the intensity of light transmitted through the analyser when $\theta = 60°$.

Solution

a) Polarized light has the direction of the electric field always in the same plane.

b) $I = I_0 \cos^2 \theta = I_0 \cos^2 60 = \dfrac{I_0}{4}$

> **» Assessment tip**
>
> It is easy to forget the cos^2 in Malus's law. Remember that the component of wave amplitude in the final beam is $A_0 \cos \theta$ and $I \propto A^2$, so $I \propto A_0^2 \cos^2 \theta$.

a) Radio waves are emitted by a straight conducting rod antenna (aerial). The plane of polarization of these waves is parallel to the transmitting antenna.

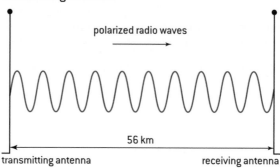

polarized radio waves

56 km

transmitting antenna   receiving antenna

An identical antenna is used for reception. Suggest why the receiving antenna needs to be parallel to the transmitting antenna.   [2]

This answer could have achieved 1/2 marks:

> ▲ The answer implies that a maximum is required in 'all the radio waves will go through'.

> ▼ The answer does not address the issue that the receiving antenna must be aligned with plane of the polarization and that this will be parallel to the transmitting antenna.

The receiving antenna needs to be parallel to the transmitting antenna because we need to make sure the coming radio waves is at an angle of 0° or 180° to the receiving antenna, so all the radio waves will go through and won't be Polarized.

b) The receiving antenna becomes misaligned by 30° to its original position.

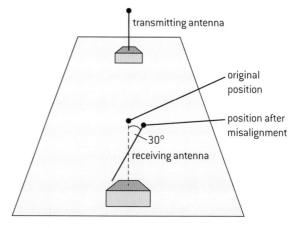

transmitting antenna

original position

position after misalignment

30°

receiving antenna

The power of the received signal in this new position is 12 μW.

Calculate the power that was received in the original position.   [2]

This answer could have achieved 0/2 marks:

> ▼ There is a suggestion in the answer that the component depends on the angle rather than the cosine of the angle. There is no attempt to use Malus's law and so no credit can be gained.

$90° \rightarrow 0W$

$30°$ is $\frac{1}{3}$ of that

$12 \mu W \times 3 = 36 \mu W$

4.4 ` WAVE BEHAVIOUR

You must know:

✔ the meaning of reflection, refraction, Snell's law, critical angle and total internal reflection

✔ an experimental method for determining refractive index

✔ the meaning of diffraction and that it occurs with all wave types

✔ the meaning of the path difference between two waves

✔ the details of a double-slit interference pattern.

You should be able to:

✔ solve problems involving reflection, refraction, Snell's law, critical angle and total internal reflection

✔ interpret diagrams showing waves at the boundary of two media

✔ describe the diffraction pattern formed when plane waves are normally incident on a single slit

✔ describe the intensity pattern produced in double-slit interference.

When waves are incident on the interface between two different media, changes in direction of the wave are observed.

- *Reflection*—the wave continues to travel in the original medium.

- *Refraction*—the wave travels in the new medium.

The angles of incidence, reflection and refraction are always measured from the normal to the interface between the media. Incident, reflected and refracted rays all lie in the same plane.

The ratio $\dfrac{n_1}{n_2}\left(=\dfrac{c_1}{c_2}\right)$ is the (relative) refractive index going from

medium 1 to medium 2 and is conveniently written as $_1n_2$. The

absolute refractive index is $\dfrac{\text{speed of light in a vacuum}}{\text{speed of light in the medium}}$ but, as

the speed in air is close to that in a vacuum, in practice, the relative refractive index going from air to the medium is used.

Media with large *optical densities* have large values of n and correspondingly small wave speeds.

Rays are reversible: when waves are reversed, their rays trace out the same path but in the opposite direction.

When waves travel between media from an optically more dense to a less dense medium, the angle of refraction is greater than the angle of incidence (Figure 4.4.2).

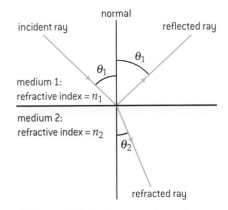

▲ Figure 4.4.1. Rays at a boundary between two media

The **reflected angle** is always equal to the **incidence angle** θ_1 when measured from the normal to the interface between the media.

The **refraction angle** θ_2 is related to the incidence angle by
$$\frac{\sin\theta_1}{\sin\theta_2}=\frac{n_1}{n_2}=\frac{c_1}{c_2},$$
where n is the absolute refractive index of the medium and c is the speed of the wave in the medium. This is **Snell's law**.

>> Assessment tip

When a wave goes from medium 1 through medium 2 to medium 3, then $_1n_3 = {_1n_2} \times {_2n_3}$. Expressions such as this can be useful when the rays travel through more than one medium.

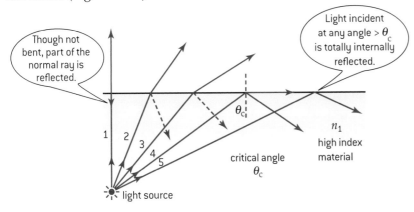

▲ Figure 4.4.2. Critical angle and total internal reflection

The **critical angle** c occurs when incident ray θ_1 gives rise to a refracted ray with $\theta_2 = 90°$. So

$$\frac{\sin c}{\sin 90} = {}_2n_1 = \frac{1}{{}_1n_2} \text{ and } \sin c = \frac{1}{{}_1n_2}$$

since $\sin 90° = 1$. Take care with the n subscripts here.

Total internal reflection occurs when waves travel from a medium with high refractive index to one with a lower refractive index, and the angle of incidence is greater than the critical angle.

As the angle of incidence increases from a small value, there comes a point where the angle of refraction grazes the interface—the incident angle is the *critical angle*. A weak reflected ray is observed for these angles. For larger incident angles, there is no refraction; only a strong *total internal reflection* is seen.

Example 4.4.1

A plane mirror consists of a parallel-sided glass sheet coated with a reflector. A ray of light is incident from air above the glass and enters the glass at an angle of 30° to the glass surface.

The refractive index of the glass is 1.5.

a) Determine the angle of reflection at the reflector.

b) Calculate the critical angle for the glass–air interface.

Solution

a) The angle of incidence at the air–glass interface is 60° and Snell's law shows that:

$\sin r = \dfrac{\sin 60}{1.5}$. Hence, r is 35°. Because the glass has parallel sides, the angle of incidence at the glass–mirror interface is also 35°.

b) Using $\sin c = \dfrac{1}{{}_1n_2}$ gives c = $\sin^{-1} = \dfrac{2}{3}$ and c = 42°.

Waves incident on a slit or an obstacle undergo **diffraction**. The wave 'spreads out' as it interacts with the aperture (Figure 4.4.3).

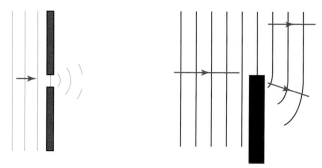

▲ Figure 4.4.3. Diffraction at a slit and around an edge

The effect is most obvious when the aperture dimensions are of the same order as or are smaller than the wavelength. Diffraction is easily demonstrated with light but is observed for all waves. For example, long-wavelength radio waves are diffracted by hills to allow radio reception in what is apparently a shadow area.

>> Assessment tip

Take care drawing the single-slit diffraction pattern. Remember:

* the central maximum is about nine times more intense than the first maximum

* maxima further from the central position have even smaller intensities

* maxima are not symmetrical.

>> Assessment tip

Take care not to confuse refraction and diffraction. It is easy to write one instead of the other when under exam pressure so be careful to avoid this error.

Example 4.4.2

Monochromatic light is incident on a narrow slit.

Sketch a graph showing the variation of light intensity with distance from the centre of the diffraction pattern.

Solution

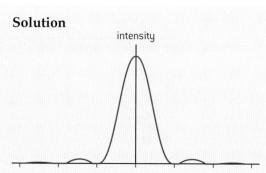

When light of a single wavelength (*monochromatic* meaning one colour) is diffracted by a single slit, the red light is diffracted more than the blue light.

When a wave travels through two parallel slits, the diffracted beams interact. This produces double-slit interference and gives rise to a distinctive fringe pattern (Figure 4.4.4).

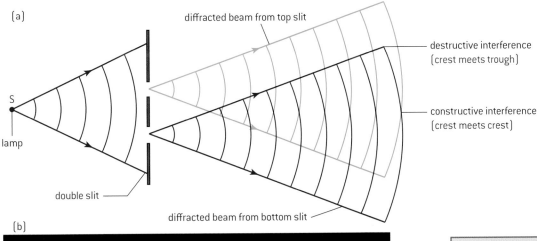

(a)

diffracted beam from top slit

destructive interference (crest meets trough)

constructive interference (crest meets crest)

S

lamp

double slit

diffracted beam from bottom slit

(b)

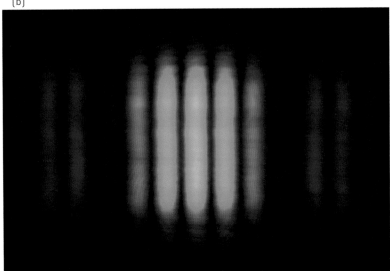

> The **double-slit interference equation** $s = \dfrac{\lambda D}{d}$ gives the separation s of successive bright fringes when light of wavelength λ is incident on two slits separated by distance d and the screen is placed a distance D from the slits.
>
> For **constructive interference** two rays must have path difference $n\lambda$.
>
> For **destructive interference** two rays must have path difference $\left(n+\dfrac{1}{2}\right)\lambda$, where n is an integer.

▲ **Figure 4.4.4.** (a) Interference of two diffracted beams from double slits and (b) the appearance of the pattern

The maxima (bright fringes) of this pattern are caused by two waves arriving at the screen *in phase*. The two waves then superpose as described in Topic 4.3. The dark fringes (minima) occur when the waves arrive 180° (or π) *out of phase*.

Example 4.4.3

Two identical loudspeakers are placed 0.75 m apart. Each loudspeaker emits sound of frequency 2.0 kHz. Y is midway between the loudspeakers and 5.0 m away from the line that joins them. Y is a position of maximum sound intensity.

Positions X and Z are equidistant from Y and have zero sound intensity. There are further maxima and minima beyond X and Z.

The speed of sound in air is 340 m s⁻¹.

a) Explain why the sound is:

 i) a maximum at Y

 ii) a minimum at X and Z.

> **>> Assessment tip**
>
> Take care with the phrase **out of phase**. This describes two waves that have **any** phase difference, other than zero, and is ambiguous. Always quote the phase difference as well, for example, $\dfrac{\pi}{2}$ rad out of phase.

b) Calculate:

    i) the wavelength of the sound

    ii) the distance XY.

Solution

a) i) Position Y is equidistant from the loudspeakers, so both waves travel an equal distance to Y. When the waves superpose, they are in phase. This produces constructive interference and an intensity maximum.

    ii) At X and Z, the waves arrive 180° out of phase; the path difference is $\left(n+\dfrac{1}{2}\right)\lambda$.

    They interfere destructively to produce complete cancellation of the superposed waves.

b) i) $\lambda = \dfrac{330}{2 \times 10^3} = 0.165$ m

   ii) The separation between maxima $= \dfrac{\lambda D}{d} = \dfrac{0.165 \times 5}{0.75} = 1.10$ m.
which is the separation between minima. XY = 0.55 m.

SAMPLE STUDENT ANSWER

A student investigates how light can be used to measure the speed of a toy train.

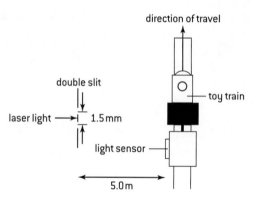

direction of travel

double slit

laser light → 1.5 mm

toy train

light sensor

5.0 m

not to scale

Light from a laser is incident on a double slit. The light from the slits is detected by a light sensor attached to the train.

The graph shows the variation with time of the output voltage from the light sensor as the train moves parallel to the slits. The output voltage is proportional to the intensity of light incident on the sensor.

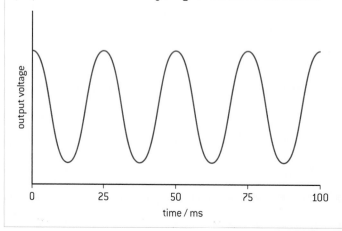

a) Explain, with reference to the light passing through the slits, why a series of voltage peaks occurs. [3]

This answer could have achieved 3/3 marks:

Light hits the two slits and diffracts into two radially propagating waves. If the path difference at the sensor is a multiple of the wavelength of the light then by superposition a maximum occurs (constructive interference). If $\left(n + \dfrac{1}{2}\right)$ multiple occur destructive interference occurs (the troughs). A series of peaks are created because they are at multiples of the wavelength. Because the voltage is proportional to intensity which is proportional to (amplitude)², the constructive interference form voltage peaks.

▲ This is a complete answer from a student who has read the question thoroughly. The process of superposition is well described and the distinction between constructive and destructive interference is clear. The answer also links intensity of the fringe pattern to the voltage peaks which is also a clear requirement of the question.

This answer could have achieved 0/3 marks:

Because as the light passing through the slits, maximum and minimum will appear as light sensor is in the different position. The direction of energy transfer in this case is parallel to the direction of particle movements in the medium.

▼ Although there is a recognition that there are maxima and minima, this really just restates the details of the graph in the question. There is no attempt to explain how the fringes form or how their intensity variation relates to changes in voltage shown on the graph.

b) The slits are separated by 1.5 mm and the laser light has a wavelength of 6.3×10^{-7} m. The slits are 5.0 m from the train track. Calculate the separation between two adjacent positions of the train when the output voltage is at a maximum. [1]

This answer could have achieved 1/1 marks:

$$S = \frac{\lambda D}{d} = \frac{(6.3 \times 10^{-7})(5)}{(1.5 \times 10^{-3})} = 2.1 \times 10^{-3}\,m = 2.1\,mm$$

▲ A careful, legible answer with a clear substitution and evaluation.

4.5 STANDING WAVES

You must know:

✔ the nature of standing waves

✔ how nodes and antinodes form

✔ how boundary conditions influence the node–antinode pattern in a standing wave

✔ the use of the term harmonic.

You should be able to:

✔ describe the nature and formation of standing waves in terms of superposition

✔ distinguish between standing and travelling waves

✔ observe, sketch and interpret standing wave patterns in strings and pipes

✔ solve problems involving the frequency of a harmonic, length of the standing wave and the wave speed.

Standing waves are constant in space but vary with time. They arise from the superposition of two waves travelling in opposite directions.

Permanent zero positions on a standing wave are **nodes**; the peak amplitude position is an **antinode**.

This table shows comparisons between standing and travelling waves:

	Standing wave	Travelling wave
amplitude	Zero at nodes; maximum at antinodes equal to $2x_0$	Same for all particles in wave amplitude is x_0
energy	No energy transfer but there is energy associated with the particle motion	Energy transfer
frequency	Same for all particles except those at node (at rest)	Same for all particles
phase	Phase for all particles between adjacent nodes is the same. Phase difference of π rad between one internodal segment and the next	All particles within a wavelength have different phases (difference varying from 0 to 2π rad)
wavelength	$2 \times$ distance between any pair of adjacent nodes or antinodes	Distance between nearest particles with the same phase

The formation of a standing wave is closely linked to resonance, which is described in Option B.4.

Standing waves form when a wave is reflected at a boundary and superposes with the original wave. Figure 4.5.1 shows the cases possible when a wave travelling along a string meets an interface with another string of greater or lesser density. The wave speed in a string depends on the mass of the string per unit length: the larger this quantity, the smaller the wave speed.

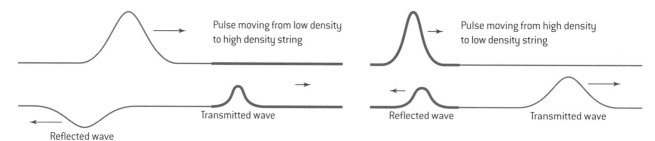

Pulse moving from low density to high density string

Transmitted wave

Reflected wave

Pulse moving from high density to low density string

Reflected wave   Transmitted wave

▲ Figure 4.5.1. Pulses moving towards free and fixed ends of strings

The behaviour of the linked strings should remind you of the behaviour of light rays travelling between two media from Topic 4.4. There are also links to Topic 9.3, where thin-film interference obeys similar rules. Use these links to help your understanding.

When the end of the string is fixed or free, Figure 4.5.1 still applies. When free, the reflected pulse is the original amplitude non-inverted; when fixed, the reflected pulse is inverted. There is no transmitted pulse is these cases.

Notice that:

- the phase of the reflected wave depends on the nature of the interface

- there is both a transmitted and reflected wave when the string is linked to another.

The boundary fixes the standing-wave shape for a specific frequency. Figure 4.5.2 shows what happens when a string is stretched between two fixed points and a wave is superposed with its reflection. At frequency f_1, a standing-wave forms with a peak amplitude in the centre of the string. There are nodes at the fixed ends. This is the *first harmonic* and is an important oscillation mode for many stringed instruments. When there are further nodes on the string, higher harmonics are produced.

If the length of the string is L, you get the following **harmonic series**:

First harmonic, $L = \dfrac{\lambda}{2}$, $f_1 = \dfrac{c}{2L}$

$N = 1$

Second harmonic, $L = \lambda$, $f_2 = \dfrac{c}{L} = 2f_1$

$N = 2$

... and so on.

$N = 3$

▲ **Figure 4.5.2.** Formation of standing wave on a string fixed at both ends

Example 4.5.1

A standing wave oscillates with three loops on a string of length 0.78 m. The frequency of oscillation is 140 Hz.

a) Calculate the speed of transverse waves along the wire.

b) Calculate the first harmonic frequency of the wire.

Solution

a) The wavelength of the wave is $\dfrac{2 \times 0.78}{3} = 0.52$ m. As the

frequency is 140 Hz, the wave speed is $f\lambda = 140 \times 0.52 = 73$ m s^{-1}.

b) The wavelength will now be $0.78 \times 2 = 1.56$ m. f_1 is 47 Hz.

Example 4.5.2

A horizontal glass tube contains fine powder. A loudspeaker at one end of a horizontal glass tube emits a single frequency and a standing wave forms in the tube. Powder heaps occur at nodes.

Speed of sound waves in air = 340 m s^{-1}

a) Identify the type of wave formed in the tube.

b) Determine, for the sound in the tube:

  i) its wavelength

  ii) its frequency.

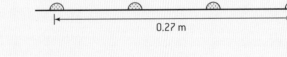

c) P and Q are points in the tube.

  Compare the frequency, amplitude and phase of air particles at P with those at Q.

Solution

a) Longitudinal because this is a sound wave in a gas.

b) i) $\lambda \left(= \dfrac{0.270}{3} \times 2 \right) = 0.18$ m

   ii) $f \left(= \dfrac{c}{\lambda} \right) = \dfrac{340}{0.18} = 1.9$ kHz

c) The frequencies are the same. P is midway between two nodes so must be an antinode.

  Q is between the antinode and node; so, the amplitude of P is greater than the amplitude of Q. P and Q are in adjacent node–node segments. This means that the wave displacements are in opposite directions; so, the phase difference is π rad.

Standing waves also form in the air in *pipes*. These lead to the series of harmonics shown in Figure 4.5.3.

In **pipes**, the nodes and antinodes are formed at points of zero amplitude and peaks, respectively, as for strings.

A closed end is a **displacement node**; an open end is a **displacement antinode**.

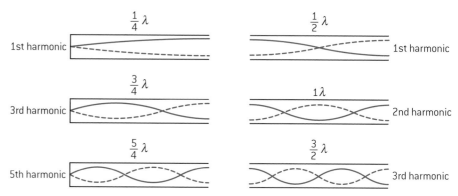

▲ Figure 4.5.3. Standing waves in pipes open at one end and open at both ends

The phase of the reflections at the ends of a pipe arises from the nature of the boundary. Air molecules at a closed end cannot move along the axis so an incident wave must be reflected as an inversion of the original to ensure that the molecules remain stationary. At an open end, the pressure must always be atmospheric, and this is therefore a pressure node. Topic 4.2 showed that where the pressure is atmospheric there is a displacement maximum. There must be a maximum amplitude of the air molecules at the open end.

Example 4.5.3

An organ pipe of length L is closed at one end.

a) Sketch the displacement of the air in the pipe when the emitted frequency of pipe is:

  i) the lowest possible f_1

  ii) the next harmonic above the lowest possible f_2.

b) Deduce an expression for $\dfrac{f_1}{f_2}$.

Solution

a) i) and ii) see Figure 4.5.3 (left-hand column, first and third harmonics).

b) L is the length of the pipe.

    So $\lambda_1 = 4L$ and $\lambda_2 = \dfrac{4L}{3}$.

    Use $c = f\lambda$ and $f_1 = \dfrac{c}{4L}$ and $f_2 = \dfrac{3c}{4L}$.

    So $\dfrac{f_1}{f_2} = \dfrac{1}{3}$.

In another experiment the student replaces the light sensor with a sound sensor. The train travels away from a loudspeaker that is emitting sound waves of constant amplitude and frequency towards a reflecting barrier.

The sound sensor gives a graph of the variation of output voltage with time along the track that is similar in shape to the graph on page 46. Explain how this effect arises. [2]

This answer could have achieved 0/2 marks:

The sound particle is emitting further apart, as the more particles moves to sound sensor, the higher amplitude appear, when they move further, the lowest point appear, when the sound reflect from reflecting barrier, new maximum point appear.

▼ The key to answering this question is to recognise that a standing wave forms between the barrier and the loudspeaker. A reference to the reflection at the barrier is required. This needs to go on to describe the formation of the standing wave and its sequence of nodes and antinodes. There are a number of separate points that can be made and any two relevant points will be given full credit.

Practice problems for Topic 4

Problem 1
An object performs simple harmonic motion with displacement x_0 and time period T.

a) Identify the phase difference between velocity and displacement for this motion.

b) Sketch a graph to show the variation of kinetic energy with time for the object for time T. Explain your answer.

Problem 2
A particle oscillates with simple harmonic motion without any loss of energy. What is true about the acceleration of the particle?

A It is always in the opposite direction to its velocity.

B It is least when the speed is greatest.

C It is proportional to the frequency.

D It decreases as the potential energy increases.

Problem 3
For a sound wave travelling through air, explain what is meant by *particle displacement, amplitude* and *wavelength*.

Problem 4
A transverse wave travels along a string. Each point in the string moves with simple harmonic motion.

a) Sketch a graph showing

(i) the variation of displacement for a point on the wave with time

(ii) the variation of displacement with distance along the wave.

Label your graphs and axes clearly.

b) Label, where possible, on your graphs

(i) the amplitude x_0 of the wave

(ii) the period T of the vibrations

(iii) the wavelength λ of the wave

(iv) points, P and Q, which have a phase difference of $\frac{\pi}{2}$.

Problem 5
The frequency range that a girl can hear is from 30 Hz to 16 500 Hz. The speed of sound in air is $330 \, \text{m s}^{-1}$. Calculate the shortest wavelength of sound in air that the girl can hear.

Problem 6
Plane-polarized light is incident on a polarizing filter that can rotate in a plane perpendicular to the direction of the light beam.

When the incident intensity is $32 \, \text{W m}^{-2}$, the transmitted intensity is $8.0 \, \text{W m}^{-2}$.

Calculate the transmitted intensity when the polarizer has been rotated through 90°.

5 ELECTRICITY AND MAGNETISM

5.1 ELECTRIC FIELDS

You must know:

✔ the nature of electric charge

✔ what is meant by electric field

✔ Coulomb's law

✔ that an electric current is movement of charge carriers

✔ what is meant by direct current

✔ what is meant by potential difference.

You should be able to:

✔ identify two forms of charge and the direction of forces between them

✔ identify the sign and nature of charge carriers in a metal

✔ identify the drift speed of charge carriers

✔ solve problems involving electric fields and Coulomb's law, the drift-speed equation, current, potential difference and charge.

🔗 The electric field concept is developed in Topic 10.1.

When there is an electric field acting at a point, a charge placed at that point will have a field acting on it. If this charge is free to move, then it will be accelerated.

Electric field strength E is a vector. It acts in the direction of the force acting on a positive charge.

$$E = \frac{\text{force acting on a positive point test charge}}{\text{magnitude of the charge}} = \frac{F}{q}$$

The unit of E is newtons per coulomb ($N\,C^{-1}$). In fundamental units, this is $kg\,m\,s^{-3}\,A^{-1}$.

The definition involves a 'point test charge', this is taken to be a charge so small that it does not disturb the field.

Electric current is the movement of charge. Its unit is the ampère (A)—an SI fundamental unit. It is defined in terms of magnetic effects.

Electric charge is defined in terms of the ampère. The current is 1 A when one coulomb of charge flows past a point in 1 s. Its unit is coulomb (C). In fundamental units, this is 1 A s.

$$I = \frac{\Delta q}{\Delta t}$$ where I is the electric current, q is the charge that flows, and t is the time.

When the current is constant, $q = It$.

Drift speed v of the charge carriers is related to the density of carriers n the charge on each carrier q the cross-sectional area of the carrier A and the electric current I.

$$v = \frac{I}{nqA}$$

Density of the carriers is the number of carriers in each cubic metre of the conductor.

Electric currents can exist in solids, liquids and gases.

Conduction in solids is usually due to the movement of one type of charge carrier. Solid conductors (metals, and materials such as carbon) contain fixed positive ions that make up the bulk of the material. These ions release electrons to a 'sea' of free electrons as part of the chemical bonding.

The electrons are accelerated in the electric field. Kinetic energy is transferred to them from the energy source (an electric cell, battery or power supply). Electrons then collide with the fixed positive ions.

Because the charge carriers make repeated collisions with the ions, they gain and lose kinetic energy as they travel. This leads to an average speed for the carriers known as the *drift speed*.

Overall, the field transfers energy to the fixed ions from the energy source.

Coulomb's law describes how the force between two point charges varies with their separation r:

$F = \dfrac{kq_1q_2}{r^2}$, where q_1 and q_2 are the magnitudes of the point charges

(+ or −). k is the constant of proportionality (its magnitude depends on the units used for distance and charge).

> **Constant of proportionality, k** is $8.99 \times 10^9\,\mathrm{N\,m^2\,C^{-2}}$ (measuring charge in coulombs, force in newtons, and distance in metres).
>
> **Permittivity of free space, ε_0** is $8.85 \times 10^{-12}\,\mathrm{F\,m^{-1}}$ (F is the farad; see Topic 11.3).
>
> To rationalise electric and magnetic units in SI units, $k = \dfrac{1}{4\pi\varepsilon_0}$.
>
> Coulomb's law $F = \dfrac{q_1q_2}{4\pi\varepsilon_0 r^2}$ when the charges are in a vacuum (or in air, which has a permittivity almost equal to that of a vacuum).
>
> When a medium other than a vacuum or air is in use, ε_0 is replaced by $\varepsilon_0\varepsilon_r$ where ε_r is the **relative permittivity** of the medium.

When both charges are positive, the force in Coulomb's law is also positive. The forces repel. When one charge is positive and the other charge is negative, the force is negative. The forces attract.

Example 5.1.1

Two identical spheres have charges of magnitude q and $2q$. When separated, the attractive force between them is F. The spheres touch to share charge and are then returned to their original separation.

Calculate the force between the charges after the separation.

Solution

Call the original separation d. $F = \dfrac{kq \times 2q}{d^2} = \dfrac{2kq^2}{d^2}$

This is an attractive force, so one of the charges must be positive and the other must be negative.

When they combine, therefore, the total remaining charge is q.

This is shared between the two spheres, $\dfrac{q}{2}$ each. The new force is

$F' = \dfrac{k\left(\dfrac{q}{2}\right)^2}{d^2} = \dfrac{kq^2}{4d^2}$, which is repulsive and equal to $\dfrac{F}{8}$.

When charges move, energy is transferred to them from the source. This energy is called *potential difference*. It is measured in terms of the energy transferred per unit of charge.

Assessment tip

Assigning the direction of **conventional current** needs care. Early scientists labelled the charge carriers in metals as positive. We now know that charge carriers in metals are electrons and are negative. Conventional current is due to the flow of **positive** charges; direction rules rely on this.

Unless you are told otherwise, always assume that 'current' refers to a conventional current.

Electric charge exists in **positive** and **negative** forms.

Like charges **repel**. Unlike charges **attract**. A charged body can also attract an uncharged body due to the charge movement within the neutral object.

Potential difference V is measured in volts (V). One volt is the energy W transferred between two points when one coulomb of charge q moves between them.

So $V = \dfrac{W}{q}$ and $1\,\mathrm{V} \equiv 1\,\mathrm{J\,C^{-1}}$.

This means that electric field strength $E\left(= \dfrac{F}{q}\right)$ can be written as

$\dfrac{F \times \text{distance}}{q \times \text{distance}} = \dfrac{\text{energy}}{q} \times \dfrac{1}{\text{distance}}$

$= \dfrac{\text{pd}}{\text{distance}}$.

So $E = \dfrac{F}{q} = \dfrac{V}{d}$.

This alternative way to express E is the subject of Topic 10.2.

>> **Assessment tip**

Example 5.1.2 part a) is a 'show that' question. You must make every step clear **including the final calculation**.

Example 5.1.2

A charged point sphere of mass 2.1×10^{-4} kg is suspended from an insulating thread between two vertical parallel plates, 80 mm apart.

A potential difference d of 5.6 kV is applied between the plates.

The thread then makes an angle of 8.0° to the vertical.

a) Show that the electrostatic force F on the sphere is given by 0.29 mN.

b) Calculate:

   i) the electric field strength between the plates

   ii) the charge on the sphere.

Solution

a) Horizontally, $F = T \sin 8$, vertically, $mg = T \cos 8$
$$F = mg \tan 8 = 2.1 \times 10^{-4} \times 9.81 \times \tan 8 = 0.29 \text{ mN}$$

b) i) $E = \dfrac{V}{d}$ gives $\dfrac{5600}{80 \times 10^{-3}} = 70 \text{ kV m}^{-1}$

   ii) $q = \dfrac{F}{E}$ gives $\dfrac{2.9 \times 10^{-4}}{7.0 \times 10^{4}} = 4.1 \times 10^{-9} \text{ C}$

SAMPLE STUDENT ANSWER

a) An electric cable contains copper wires covered by an insulator. An electric field exists across the ends of the cable. Discuss, in terms of charge carriers, why there is a significant current only in the copper. [3]

This answer could have achieved 1/3 marks:

Insulator made by specific material, which the electron cannot pass through hence no electric current. However, copper wires can carry large number of free-moving electrons, which have very good electron conductivity. Hence, there is significant electric current in copper wires.

▲ The answer includes the idea that conductors have high numbers of charge carriers, whereas insulators do not (by implication).

▼ There is no statement that the charge carriers in the conductor are electrons. There is no link between the number of charge carriers and the electric field mentioned in the question. This is clearly required as the question states that both conductor and insulator have the same electric field and, therefore, potential difference across them. The electrons are accelerated by the field and it is this motion that gives rise to the current.

This answer could have achieved 3/3 marks:

Copper wire is a conductor which means it has delocalized or free electrons (the charge carriers) which can move. The electric field exerts a force on them making them move which is the definition of electric current. The insulators electrons are not free to move, and without many free moving charge carriers (electrons) the electric current will not be significant.

▲ All three marking points are here. 'Accelerate' would have been better than 'move', but the sense is correct.

b) A wire in the cable has a radius of 1.2 mm and the current in it is 3.5 A. The number of electron s per unit volume of the wire is 2.4×10^{28} m³. Show that the drift speed of the electrons in the wire is 2.0×10^{-4} ms⁻¹. [1]

This answer could have achieved 0/1 marks:

▼ The value for q (which should be 1.6×10^{-19} C) seems to be replaced by N_A (the Avogadro number – see Topic 3.3) which is incorrect. Furthermore, there is no evidence that this calculation was actually carried out by the student.

$$v = \frac{I}{nAq} = \frac{3.5}{2.4 \times 10^{28} \times (1.20 \times 10^{-3})^2 \, 12 \times Na}$$

5.2 HEATING EFFECT OF AN ELECTRIC CURRENT

You must know:

✔ Kirchhoff's circuit laws

✔ what is meant by the heating effect of an electric current and its consequences

✔ that resistance is expressed as $R = \dfrac{V}{I}$

✔ Ohm's law

✔ the definition of resistivity

✔ the definition of power dissipation.

✔ what is meant by ideal and non-ideal ammeters and voltmeters

✔ practical uses of potential divider circuits, including the advantages of a potential divider over a series resistor in controlling a simple circuit.

You should be able to:

✔ draw and interpret circuit diagrams

✔ identify ohmic and non-ohmic components by considering V–I characteristic graphs

✔ solve problems involving potential difference, current, charge, Kirchhoff's circuit laws, power, resistance and resistivity

✔ investigate the combined resistance of arrangements of series and parallel resistors

✔ investigate experimentally and describe the factors that affect the resistance of a conductor.

An electric current causes the transfer of thermal energy when charge flows in a component. An electric current can also cause chemical and magnetic effects.

Electrical power dissipated is the rate with which energy is transferred. Remember that the potential difference V across the component is the energy W transferred per unit charge. The power P dissipated in a component in time t is $\dfrac{W}{t}$.

But $V = \dfrac{W}{q}$ and $q = It$. So $V = \dfrac{W}{It}$ and $IV = \dfrac{W}{t} = P$.

Using the definition of resistance $R = \dfrac{V}{I}$ leads to $P = IV = I^2R = \dfrac{V^2}{R}$.

These are important equations—you should try to memorise them.

The *electrical resistance* of a component indicates the difficulty that charges have when moving through the component—or, alternatively, the ease with which energy can be transferred from charge carriers to the component.

> **Resistance**
>
> $$R = \frac{\text{potential difference across a component}}{\text{current in the component}} = \frac{V}{I}$$
>
> Resistance is measured in ohms (Ω). $1\,\Omega$ is the resistance of a component that has a potential difference of $1\,V$ across it when the current through it is $1\,A$.
>
> The resistance of a component depends on its size, shape and the material from which it is made. Experiments show that resistance R is:
>
> • proportional to length l
>
> • proportional to the resistivity of the material ρ, which is a shape-independent quantity
>
> • inversely proportional to area A

Example 5.2.1

The 'lead' in a pencil is a conductor. It is made from material which has a resistivity of $4.0 \times 10^{-3}\,\Omega\,m$.

Determine the resistance of a pencil 'lead' of length $80\,mm$ and diameter $1.4\,mm$.

Solution

$R = \dfrac{\rho l}{A} = \dfrac{4.0 \times 10^{-3} \times 80 \times 10^{-3}}{\pi \times (0.70 \times 10^{-3})^2} = 0.20\,k\Omega$ to 2 significant figures.

> **Resistivity** $\rho = \dfrac{RA}{l}$. The unit of resistivity is the $\Omega\,m$. All samples of the same pure material have the same resistivity.

Example 5.2.2

A heating element wire is 4.5 m long with diameter 1.5 mm.

Its resistivity is $9.6 \times 10^{-6}\,\Omega\,m$ at its operating temperature.

It is used with a 110 V supply.

Calculate the power rating of the heating element wire.

Solution

Area of the wire $= \pi \times (0.75 \times 10^{-3})^2 = 1.77 \times 10^{-6}\,m^2$.

Substituting into the resistivity equation gives

$$R = \frac{\rho l}{A} = \frac{9.6 \times 10^{-6} \times 4.5}{1.77 \times 10^{-6}} = 24.4\,\Omega$$

The power dissipated is $P = \dfrac{V^2}{R} = \dfrac{110^2}{24.4} = 496$ W which rounds to 500 W.

A component that obeys Ohm's law is an *ohmic conductor*. Graphs of the variation of V with I for a component are called V–I characteristic graphs. They can be used to show whether a conductor is ohmic or non-ohmic.

Figure 5.2.1 shows the V–I characteristics for three different conductors:

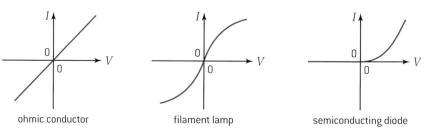

ohmic conductor   filament lamp   semiconducting diode

▲ **Figure 5.2.1.** V–I characteristics graphs

When resistances are joined in series or parallel, you can calculate the resistance of the combination using these two rules:

1 In series

Add the values

$$R = R_1 + R_2 + R_3 + \ldots$$

2 In parallel

Add the reciprocals of the values

$$\frac{1}{R} = \frac{1}{R_1} + \frac{1}{R_2} + \frac{1}{R_3} + \ldots$$

When there is a combination of series and parallel, calculate the parallel part of the network first.

Example 5.2.3

An electrical cable consists of eight parallel strands of copper wire, each of diameter 2.5 mm.

The resistivity of copper is $1.6 \times 10^{-8}\,\Omega\,m$.

The cable carries a current of 20 A.

a) Calculate:

   i) the cross-sectional area of a **single strand** of copper wire

   ii) the resistance of a 0.10 km length of the eight-strand cable.

b) Calculate the potential difference between the ends of the eight-strand cable.

c) State **one** advantage of using a stranded cable rather than a solid core cable with copper of the same total cross-sectional area.

Solution

a) i) The area of one strand is $\pi \times (1.25 \times 10^{-3})^2 = 4.91 \times 10^{-6}\,\text{m}^2$.

ii) The resistance of one strand is $\dfrac{1.6 \times 10^{-8} \times 100}{4.91 \times 10^{-6}} = 0.326\ \Omega$.

The resistance of the cable is $\dfrac{1}{8}$ of this because they are in parallel: $0.0407\ \Omega$.

b) $V = IR = 20 \times 0.0407 = 0.81$ V

c) Possible answers include: the flexibility of a cable compared with one strand; the cable will still conduct even if one strand breaks; the larger surface area gives a better heat dissipation.

Some rules follow from the conservation of charge and energy.

Figure 5.2.2 shows the currents in a circuit that lead into a junction of electrical leads. These currents represent charge (and, therefore, electrons) flowing to and from the junction. If charge builds up here, it prevents further flow. The sum of the currents at the junction must be zero – having taken into account their signs.

This is *Kirchhoff's first circuit law*: $\Sigma I = 0$ (junction)

For an electrical loop, the energy per coulomb transferred into the loop by emf sources is equal to the energy per coulomb transferred out of the loop by energy sinks.

This is *Kirchhoff's second circuit law*: $\Sigma V = 0$ (loop)

Sources of energy are represented by the emf of a cell (see Topic 5.3) and sinks by the resistors. Σ sources of emf $-\Sigma IR$ for each resistor $= 0$.

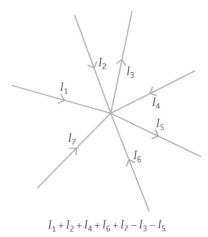

$$I_1 + I_2 + I_4 + I_6 + I_7 - I_3 - I_5$$

▲ **Figure 5.2.2.** Currents leading to junction in a circuit, and the sum of the currents

Example 5.2.4

Determine currents I_1 and I_2 for this circuit.

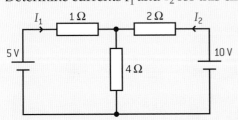

Solution

First, label the junctions, currents and emfs.

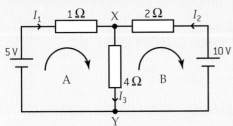

At junction X: $I_1 + I_2 = I_3$ (Kirchhoff's first circuit law; Y is similar but with opposite signs).

Voltmeters are placed in parallel with a component to measure the potential difference across it.

Ideal voltmeters have infinite resistance (otherwise the current in them would affect the circuit).

Ammeters are placed in series to measure the current in a component.

Ideal ammeters have zero resistance and do not dissipate any energy.

>> Assessment tip

Be prepared to answer both qualitative and quantitative questions about potential dividers. You should be able to show that, for the circuit below,

$$V_1 = \frac{R_1}{R_1 + R_2} V_{\text{supply}} \text{ and}$$

$$V_2 = \frac{R_2}{R_1 + R_2} V_{\text{supply}}$$

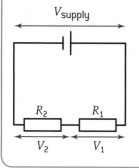

For loop A: $5 = 1 \times I_1 + 4 \times I_3$ (Kirchhoff's second circuit law).

For loop B: $-10 = -2 \times I_2 + (-4 \times I_3)$ (Kirchhoff's second circuit law again).

The next step is to eliminate I_3:

$$5 = 1 \times I_1 + 4 \times (I_1 + I_2) \text{ and } 10 = 2 \times I_2 + 4 \times (I_1 + I_2)$$

Solving these two simultaneous equations gives

$$I_1 = -0.71\,\text{A and } I_2 = 2.1\,\text{A}$$

Potential-divider circuits have advantages over variable resistors to control current. Figure 5.2.3 shows both arrangements used to determine the V–I characteristic for a metal wire.

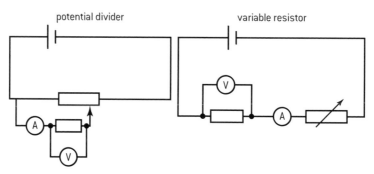

▲ **Figure 5.2.3.** V–I characteristic for a metal wire

SAMPLE STUDENT ANSWER

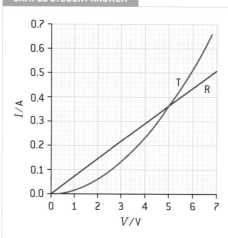

a) The graph shows how current I varies with potential difference V for a resistor R and a non-ohmic component T.

i) State how the resistance of T varies with the current going through T.   [1]

This answer could have achieved 0/1 marks:

> The resistance of T increases at a decreasing rate with the current.

▼ It is easy to get this wrong. One way to avoid this is to work out $\frac{V}{I}$ at both ends of the graph. For low current, the resistance is about $50\,\Omega$. At high current, it is about $10\,\Omega$. This would only take a few seconds on a calculator to work out.

ii) Deduce, without a numerical calculation, whether R or T has the greater resistance as $I = 0.40\,\text{A}$.   [2]

This answer could have achieved 2/2 marks:

> R has greater resistant at I = 0.40 A. As shown on the graph, R has greater voltage then T when I = 0.40 A. According to the formula, R = $\frac{V}{I}$ when V is greater it will have a greater R, hence R has greater resistance.

▲ This time, a reference is made to $\frac{V}{I}$. The fact that current is the same for both components, but V is greater for one component, gives the answer directly.

b) Components R and T are placed in a circuit. Both meters are ideal.

Slider Z of the potentiometer is moved from Y to X.

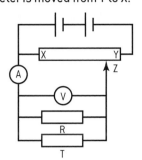

i) State what happens to the magnitude of the current in the ammeter. [1]

This answer could have achieved 0/1 marks:

> The sum of R and T's current.

▼ The focus should be on the current in the ammeter. This is a potential divider arrangement, so moving the slider to X reduces the potential difference at Z. This will also reduce the current in the ammeter (because $V = IR$, and the resistance of R and T does not change).

ii) Estimate, with an explanation, the voltmeter reading when the ammeter reads 0.20 A. [2]

This answer could have achieved 0/2 marks:

> $V = I \times R$　　$I = 0.20 A$　　$\dfrac{1}{0.3} + \dfrac{1}{0.7} = \dfrac{1}{R}$
>
> $R = \dfrac{0.06}{0.2} = 0.3\,(\Omega)$
>
> $V = 0.06\ (T)$　　$R = 0.21$　　$V = 0.21 \times 0.2 = 0.042\ (v)$
>
> $V = 0.14\ (R)$　　$R = \dfrac{0.14}{0.2} = 0.7\,(\Omega)$

▼ This approach is going to be difficult (though not impossible). Answering this question is straightforward when you remember Kirchhoff's laws. The 0.2 A is the sum of the currents in R and T. Because they are in parallel, the potential differences across them are the same.

Looking at the graph, you can see that, when $V = 2.0\,V$, the two currents are about 0.06 A and 0.14 A respectively. These add to 0.20 A and, therefore, 2.0 V is the answer.

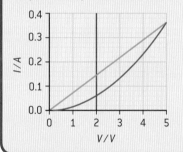

5.3　ELECTRIC CELLS

You must know:

✔ that electric cells are chemical energy stores and are sources of electromotive force (emf)

✔ the distinction between primary and secondary cells

✔ the distinction between the emf of a cell and its terminal potential difference

✔ real cells have internal resistance

✔ how to investigate practical electric cells (both primary and secondary).

You should be able to:

✔ identify the direction of charge flow required to charge a secondary cell

✔ determine the emf and internal resistance of a cell experimentally

✔ solve problems involving the emf and internal resistance and other quantities of a cell

✔ describe and interpret the variation of terminal potential difference with time as a cell discharges.

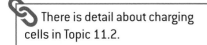 There is detail about charging cells in Topic 11.2.

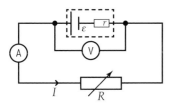

▲ **Figure 5.3.1.** A circuit to determine the internal resistance and emf of a practical cell

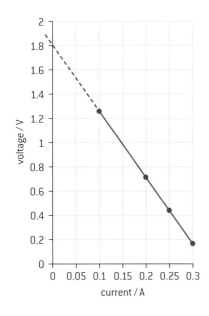

Alessandro Volta invented the first battery in 1800. This was a *primary cell* constructed from alternating copper and zinc discs separated by brine-soaked cloths. Soon, other types of primary cell and the first *secondary cell*—the lead–acid cell—were developed. Society's need for portable and compact power supplies continues to drive research into innovative secondary cells.

When a charged cell is not transferring energy, the potential difference across the cell terminals is equivalent to the *electromotive force (emf)* ε of the cell.

ε represents the maximum energy that the cell can deliver for each coulomb of charge passing through it. When transferring energy, however, a real cell always gives a smaller potential difference reading V across its terminals because energy is required to move charge through the cell itself. This loss appears as an internal resistance r of the cell.

A cell with *internal resistance* can be represented as the ideal cell plus a resistance enclosed in dotted lines (see Figure 5.3.1). For this circuit, $V = \varepsilon - IR$ or $\varepsilon = I(R + r)$. These equations follow from a Kirchhoff loop.

The equation $V = \varepsilon - IR$ provides a graphical method for determining both r (from the negative of the gradient) and ε (from the V-intercept) for a real cell. Figure 5.3.1 shows a circuit in which a variable resistor is used to control the current. As current increases, terminal potential difference across the cell drops because a higher current requires more energy to drive charge through the cell.

Example 5.3.1

The circuit in Figure 5.3.1 is used to determine the variation of V with I in the circuit. The results are given in this table.

Voltmeter reading / V	Ammeter reading / A
1.25	0.10
0.70	0.20
0.43	0.25
0.15	0.30

a) Plot a suitable graph from these data.

b) Use your graph to determine:

i) the emf of the cell

ii) the internal resistance of the cell.

Solution

a) See graph on the left.

b) i) Extrapolating the line to the y-axis gives the emf (1.8 V).

ii) r is the negative of the gradient $= -\dfrac{0.7 - 1.8}{0.2 - 0} = 5.5\,\Omega$.

A student adjusts the variable resistor and takes readings from the ammeter and voltmeter. The graph shows the variation of the voltmeter reading V with the ammeter reading I.

Use the graph to determine

a) the electromotive force (emf) of the cell. [1]

This answer could have achieved 0/1 marks:

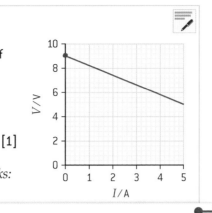

9 V

> ▼ You must clearly show the examiner what you are doing when the command is 'determine' (ideally, explain this in words). In this example, the only clue is a dot drawn at the point where the student read the graph. The clue is that there are two lines for the answer—if the examiner simply required the answer, there would have only been one.

b) the internal resistance of the cell. [2]

This answer could have achieved 1/2 marks:

Negative of the slope as $v = \varepsilon - Ir$

$$\frac{9 - 5.6}{4.2} = 0.81\,\Omega$$

> ▼ This is not a perfect answer. The student makes it clear that the negative of the (negative) slope is r, that is, a positive value. The gradient calculation is actually $\frac{5.6 - 9.0}{4.2}$ and account must be taken of the negative gradient.

5.4 MAGNETIC EFFECTS OF ELECTRIC CURRENTS

You must know:

✔ the magnetic field patterns for a bar magnet, a long straight conductor and a solenoid

✔ what is meant by magnetic force

✔ how to solve problems involving magnetic forces, fields, currents and charges.

You should be able to:

✔ sketch and interpret magnetic field patterns for long straight conductors, solenoids and bar magnets

✔ determine the direction of a magnetic field for a wire or solenoid knowing the current direction

✔ determine the direction of the force acting on a charge moving in a magnetic field and the direction of the force acting on a current-carrying conductor in a magnetic field.

Electric currents produce, and interact with, *magnetic fields*. This is the third effect of electric current. A visualization using field lines has evolved to indicate both the direction and strength of a field. Field lines also give a visual meaning to gravitational and electrostatic fields.

Although the lines are not real, properties can be assigned to them.

Magnetic field lines:

• are in the direction that a north pole would travel if free to do so

• cannot cross

• show the strength of the field— the closer lines are, the stronger the field

• start on north poles and end on south poles

• act as though they are elastic strings, trying to be as short as possible.

This table provides information about the magnetic field patterns for a bar magnet, a long straight (current carrying) wire and a solenoid. Certain rules can help you remember the field patterns.

Bar magnet

The field lines go from the north pole to the south pole—in the direction that a free north pole would travel.

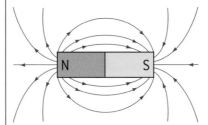

Long straight (current-carrying) wire

A pattern of circular field lines surrounding a wire.

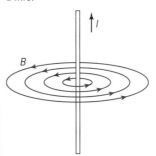

Use the right-hand rule to remember the field pattern. Imagine holding the wire with your right-hand thumb pointing in the direction of the conventional current; your fingers curl in the direction of the field.

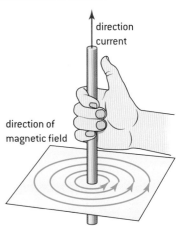

Solenoid

Field lines outside the solenoid are the same as a bar magnet. Inside, the field lines go from the south end to the north end to be continuous inside and outside.

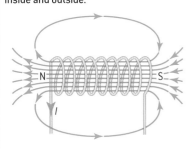

Looking into the end of the solenoid, when the conventional current circulates anti-clockwise, that is the north pole. Clockwise circulation corresponds to the south.

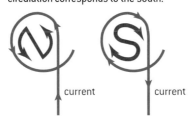

The right-hand rule can also be used.

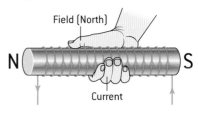

Magnetic fields occur when electric charge moves. It is not surprising that charges moving in a pre-existing magnetic field are also affected. The interaction of the moving charge and the magnetic field lead to a force acting on the charge.

The *direction of the magnetic force* is at right angles to the plane containing the magnetic field and the velocity of the charge. The force itself is proportional to:

- the velocity of the charge v
- the magnitude of the charge q
- the *strength of the magnetic field B*.

> The **direction of the magnetic force** acting on a moving charge is always at 90° to the plane that contains B and v. The equation is $F = qv\,B\sin\theta$. When the velocity of the charge and the magnetic field direction are at right angles (in other words, $\theta = 90°$ so $\sin\theta = 1$) the magnetic force acting on a moving charge is qvB.

Imagine a conductor of length l that carries a current I of charge carriers, with n charge carriers per cubic metre. The area of the conductor is A, and the charge on each carrier is q. There are nAl carriers in the segment. The total charge is $nAlq$. The total force acting on the conductor when in a magnetic field of strength B is $B(nAlq)v$, where v is the average drift speed of a charge carrier.

So $F = (nAlq)Bv = (nAvq)Bl = BIl$ because $I = nAvq$ relates drift speed to current in a conductor.

The equation $F = qvB\sin\theta$ is equivalent to $F = BIl\sin\theta$.

The force between a moving charge and a magnetic field originates in a relativistic effect. This is explored in more detail in Option A.

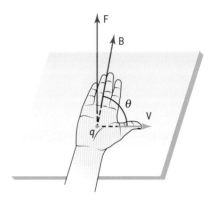

> **Magnetic field strength B** is measured in tesla (T). Rearranging the $F = qvB$ equation shows that the fundamental units for tesla are $N\,s\,m^{-1}\,A^{-1}\,s^{-1} \equiv kg\,s^{-3}\,A^{-1}$.
>
> The tesla is a large unit. The magnetic field strength of the Earth is about $50\,\mu T$.

More details about the tesla are given in Topic 11.4.

There is a link between the force on a moving charge in a magnetic field and the force that acts on a current-carrying conductor in a field. Remember that the charges in a conductor move with an average drift speed. Drift speed is considered in Topic 5.1.

>> **Assessment tip**

Rules can be used to help you with force directions. Fleming's left-hand rule relates the directions of the (conventional) current and the field to the force.

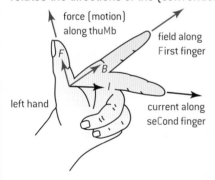

force (motion) along thuMb
field along First finger
left hand
current along seCond finger

You may have been taught other rules. Make sure that you understand how to use them – and don't use the wrong hand!

Example 5.4.1

A beta particle moves at 15 times the speed of an alpha particle through the same magnetic field. Both particles are moving at right angles to the field.

Calculate $\dfrac{\text{magnetic force on the beta particle}}{\text{magnetic force on the alpha particle}}$.

Solution

Use $F = qvB$. So $\dfrac{F_\beta}{F_\alpha} = \dfrac{Bq_\beta v_\beta}{Bq_\alpha v_\alpha} = \dfrac{e}{2e} \times 15 = 7.5$

SAMPLE STUDENT ANSWER

$I = 3.5$ A into page

wire cross-section

magnetic field

The diagram shows a cross-sectional view of a wire. The wire, which carries a current of 3.5 A into the page, is placed in a region of uniform magnetic field of flux density 0.25 T. The field is directed at right angles to the wire.

Determine the magnitude **and** direction of the magnetic force on one of the charge carriers in the wire. [2]

This answer could have achieved 2/2 marks:

$v = 2.0 \times 10^{-4} \quad I = 3.5 \quad B = 0.25 \quad q = 1.60 \times 10^{-19} \, C \quad \sin 90 = 1$

$F = qvB \sin\theta = 1.60 \times 10^{-19} \times 2.0 \times 10^{-4} \times 0.25$

$= 8.0 \times 10^{-24} \, N \text{ downwards}$

▲ The student has set out the calculation clearly and quoted a direction for the force.

Practice problems for Topic 5

Problem 1

A bicycle is powered by an electric motor that transfers energy from a rechargeable battery. When fully charged, the 12 V battery can deliver a current of 14 A for half an hour before full discharge.

a) Determine the charge stored by the battery.

b) Calculate the energy available from the battery.

c) The cyclist uses the motor when going uphill and maintains a constant speed of 7.5 m s^{-1}.

Calculate the maximum height that the cyclist can climb before the battery needs recharging.

Problem 2

A wire is 0.15 m long. A potential difference of 6.0 V is applied between the ends of the wire.

a) Calculate the acceleration of a free electron in the wire.

b) Suggest why the average speed of a free electron in the wire does not increase even though it is accelerated.

Problem 3

A 6.0 V cell and a 30 Ω resistor are connected in series. The cell has negligible internal resistance.

a) Calculate the current in the cell.

b) An arrangement of a 30 Ω resistor and a 60 Ω resistor in parallel is connected in series with the original 30 Ω resistor. Calculate the current in the cell.

Problem 4

The $I-V$ characteristic graph for two conductors A and B is shown.

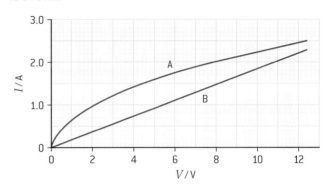

a) Explain which conductor is ohmic.

b) (i) Calculate the resistance of the conductor A when $V = 1$ V and $V = 10$ V.

(ii) A is a lamp filament. Explain why the values of resistance in part i) are different.

c) B is a wire of length 0.8 m with a uniform cross-sectional area of 6.8×10^{-8} m^2.

Determine the resistivity of B.

Problem 5

Wires **XY** and **PQ** carry currents in the same direction, as shown.

State and explain the force on XY due to PQ.

Problem 6

A wire has a cross-sectional area of 1.8×10^{-7} m^2 and contains 3.0×10^{29} free electrons per m^3.

When the potential difference across the coil is 12 V, the current in the coil is 7.2 A.

a) Calculate the mean drift velocity of electrons in the wire.

b) The current is switched on for 12 minutes.

Determine the energy transferred in the wire in this time.

6 CIRCULAR MOTION AND GRAVITATION

6.1 CIRCULAR MOTION

You must know:

✔ what is meant by angular displacement, angular velocity, period and frequency

✔ centripetal force acts on an object for it to move in a circle and that centripetal force and centripetal acceleration act towards the centre of the motion

✔ for horizontal circular motion at a constant speed, the centripetal force is constant

✔ for motion in a vertical circle at constant speed, both gravitational and centripetal forces need to be taken into account.

You should be able to:

✔ identify the origin of the centripetal force for examples of circular motion

✔ describe examples of circular motion qualitatively and quantitatively

✔ solve problems involving centripetal force, centripetal acceleration, period, frequency, angular displacement, linear speed and angular velocity.

Period T is the time taken to travel once round a circle.

Angular displacement θ is the angle through which an object moves; it is measured in degrees or radians.

Angular velocity,

$$\omega = \frac{\text{angular displacement}}{\text{time taken}} = \frac{\theta}{t},$$

where t is the time to travel θ rad. It is a vector quantity.

Angular speed has the same definition as angular velocity, but it is a scalar quantity.

The angular displacement for a complete circle is 2π rad and the periodic time is T, so $\omega = \frac{2\pi}{T}$.

Linear speed, $v = r\omega$

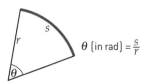

θ (in rad) $= \frac{s}{r}$

▲ Figure 6.1.1.

Quantities involved in circular motion include *period, frequency, angular displacement* and *angular velocity*.

The DP Physics course treats angular velocity as a scalar even though it can be considered as a vector with its direction along the axis of rotation.

Radian measure (rad) is used for angles. Figure 6.1.1 shows how a sector with radius r and arc length s relate to angle θ.

For a complete circle, $\theta = \frac{2\pi r}{r} = 2\pi$ rad.

The trigonometrical definitions for sine, cosine and tangent are in terms of the sides of a right-angled triangle:

$$\sin\theta = \frac{\text{opposite}}{\text{hypotenuse}}; \quad \cos\theta = \frac{\text{adjacent}}{\text{hypotenuse}}; \quad \tan\theta = \frac{\text{opposite}}{\text{adjacent}}.$$

When θ is small, $\theta \approx \sin\theta \approx \tan\theta$ because the curved arc length in the radian definition, and the opposite side in the triangle from which $\sin\theta$ is defined, are similar in length and the hypotenuse and the radius are the same.

The *angular speed* ω of an object is linked to its linear speed v.

The radian definition gives: $\theta = \frac{s}{r}$. So, $\omega = \frac{s}{r \times t} = \frac{1}{r} \times \frac{s}{t} = \frac{v}{r}$, which can be rearranged to give $v = r\omega$.

The *linear speed* is constant but its direction changes, so the linear **velocity** changes: that is, the object is accelerated. N2 tells us that velocity change is associated with a force that causes circular motion— this centripetal force is directed towards the centre of the circle.

Example 6.1.1

An astronaut is rotated horizontally at a constant speed to simulate the forces of take-off.

The centre of mass of the astronaut is 20.0 m from the rotation axis.

a) Explain why a horizontal force acts on the astronaut when the speed is constant.

b) The horizontal force acting on the astronaut is 4.5 times that of normal gravity.

Determine the speed of the astronaut.

Solution

a) Velocity is a vector quantity. The velocity of the astronaut is constantly changing because, although the speed is constant, the direction is changing. This means that there is an acceleration and, therefore, a force acting on the astronaut. This is a centripetal force.

b) $\dfrac{mv^2}{r} = 4.5\,mg$ for the centripetal force.

So $v = \sqrt{4.5gr} = \sqrt{4.5 \times 9.81 \times 20} = 30$ m s^{-1} to 2 significant figures.

> **Centripetal acceleration,**
>
> $$a = \omega^2 r = \frac{v^2}{r} = \frac{4\pi^2 r}{T^2}, \text{ where } T = \frac{2\pi}{\omega}.$$
>
> **Centripetal force** $F = mr\omega^2 = \dfrac{mv^2}{r}$,
>
> where m is the mass of an object moving at linear speed v in a circle of radius r.

> **» Assessment tip**
>
> Sometimes you will come across the term centrifugal force. This force can only arise when an observer is in a rotating frame of reference—you should **not** use this term in an examination answer; always use **centripetal force**.

Example 6.1.2

A particle P of mass 3.0 kg rotates in a vertical plane at the end of a light string of length 0.75 m.

When the particle is at the bottom of the circle, it moves with speed 6.0 m s^{-1}.

When the particle has moved through 90°, calculate:

a) the speed of the particle

b) the horizontal force on the particle.

Solution

a) The object rises by 0.75 m with a gain of gravitational potential energy of $mgh = 3 \times 9.81 \times 0.75 = 22.1$ J.

The initial kinetic energy is $\dfrac{1}{2}mv^2 = 0.5 \times 3.0 \times 6^2 = 54$ J.

Kinetic energy at 90° = 54 − 22.1 = 32 J, so speed is

$$\sqrt{\frac{2E_k}{m}} = \sqrt{\frac{2 \times 32}{3}} = 4.6 \text{ m s}^{-1}.$$

b) The centripetal force is $\dfrac{mv^2}{r} = \dfrac{3.0 \times 4.6^2}{0.75} = 85$ N

> Circular motion arises in the following common contexts.
>
> **Gravitational:** The gravitational force between a planet (or the Sun) and a satellite supplies the centripetal force that keeps the satellite in its circular orbit.
>
> **Electrostatic:** The mathematics of circular motion was used by Bohr to model the proton–electron system in the hydrogen atom.
>
> **Magnetic:** The force that acts on a charge moving in a magnetic field is at 90° to the plane containing the field direction and the velocity. This is the condition for a centripetal force.

Banking is used to make cornering easier in a vehicle.

When the vehicle corners on a horizontal surface, the friction between the tyres and the road surface must be large enough to supply the centripetal force. If the friction is not sufficient, then the vehicle skids and will attempt to go in a straight line (obeying N1).

When the surface is banked, the horizontal component of the force normal to the surface provides the centripetal force towards the centre of the circle (Figure 6.1.2).

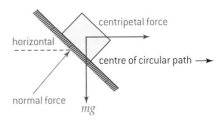

▲ **Figure 6.1.2.** Forces in cornering on a banked surface

Aircraft also bank in order to make a turn. The aircraft tilts out of the horizontal and the lift force now produces a horizontal component to the centre of the circle.

SAMPLE STUDENT ANSWER

a) A cable is wound onto a cylinder of diameter 1.2 m. Calculate the angular velocity of the cylinder when the linear speed of the cable is 27 m s^{-1}.   [2]

State an appropriate unit for your answer.

This answer could have achieved 2/2 marks:

▲ The calculation is correct and well explained. The unit is also correct.

$$\omega = \frac{v}{r} = \frac{27 \times 2\pi}{1.2 \times \pi} = \frac{27}{0.6} = 45 \text{ radianss}^{-1}$$

This answer could have achieved 0/2 marks:

▼ Although there is a reasonable start with a correct identification of the radius and quoting the correct equation, the solution becomes confused when T appears. The unit is incorrect for circular motion.

$$D = 1.2\,m \qquad r = 0.6\,m$$
$$v = \omega r = \frac{2\pi}{T} \times r = \frac{2\pi}{11} \times 0.6\,m$$
$$= 0.343\,ms^{-1}$$

6.2   NEWTON'S LAW OF GRAVITATION

You must know:

✔ Newton's law of gravitation

✔ that Newton's law of gravitation relates to point masses but can be extended to spherical masses of uniform density

✔ that the gravitational field strength at a point is the gravitational force per unit mass experienced by a point test mass placed at that point.

You should be able to:

✔ describe the relationship between gravitational force and centripetal force

✔ apply Newton's law of gravitation to the motion of an object in a circular orbit around a point mass

✔ solve problems involving gravitational force, gravitational field strength, orbital speed and orbital period of a satellite

✔ determine the gravitational field strength at a point due to two masses.

The constant of proportionality G is given by $G = \dfrac{Fr^2}{m_1 m_2}$.
In SI units, G has the magnitude $6.67 \times 10^{-11}\,N\,m^2\,kg^{-2}$ (or, in fundamental units, $m^3\,s^{-2}\,kg^{-1}$).

Newton's law of gravitation relates the gravitational force between two objects to their masses and separation. It states that the gravitational force F between two point objects of mass m_1 and m_2 is related to the distance r between the objects by

$$F \propto -\frac{m_1 \times m_2}{r^2}.$$

The negative sign indicates that the force is attractive (although you can ignore this for gravitation).

🔗 There is more detail about the gravitational interactions of objects at the planetary scale in Topic 10.

Newton's law refers to point masses but can be extended to spherical masses of uniform density. In this case, r is the distance between the centres of the masses.

Example 6.2.1

Knowledge of the acceleration of free-fall g_E at the Earth's surface is one way to determine the mass of the Earth m_E.

a) Show that $m_E = \dfrac{g_E r_E^2}{G}$, where r_E is the radius of the Earth.

b) The gravitational field strength of the Earth at its surface is 6.1 times that of the Moon at the surface of the Moon. Calculate the mass of the Moon as a fraction of the mass of the Earth.

Radius of Moon $= 1.7 \times 10^6$ m; radius of Earth $= 6.4 \times 10^6$ m

Solution

a) $mg_E = \dfrac{Gmm_E}{r_E^2}$. Cancelling and rearranging gives $r_E^2\, g_E = Gm_E$

and hence the result.

b) The equation also applies to the Moon.

$$m_E = \frac{g_E r_E^2}{G} \text{ and } m_M = \frac{g_M r_M^2}{G}.$$

Dividing gives $\dfrac{m_M}{m_E} = \dfrac{g_M r_M^2}{g_E r_E^2}$. The fraction is $\dfrac{1}{6.1} \times \left(\dfrac{1.7}{6.4}\right)^2 = 0.012$.

Gravitational field strength is the force per unit mass that acts on a mass in a gravitational field.

Example 6.2.2

The gravitational field strength at the surface of planet P is $17\,\text{N kg}^{-1}$. Planet Q has twice the diameter of P. The masses of the planets are the same.

Calculate the gravitational field strength at the surface of P.

Solution

From Newton's law of gravitation, $g \propto \dfrac{1}{r^2}$

Therefore, $\dfrac{g_Q}{g_P} = \dfrac{r_P^2}{r_Q^2}$ and $g_Q = g_P \times \dfrac{r_P^2}{r_Q^2} = \dfrac{17}{2^2} = 4.3\,\text{N kg}^{-1}$

Topic 6.1 is linked here to describe satellite orbits, since gravitational force provides the centripetal force. For a circular orbit (assumed here), the force between satellite and planet is at right angles to the direction of motion of the satellite.

Example 6.2.3

A satellite orbits above the equator so that it stays over the same point on Earth.

Calculate:

a) the angular speed of the satellite

b) the radius of the orbit of the satellite.

For an object on the Earth's surface, a mass m is gravitationally attracted to the centre of Earth, mass m_E. The law becomes

$$F = \frac{Gm \times m_E}{r_E^2}$$

where r_E is the radius of the Earth.

This is the force of gravity; the weight of the object is written as mg leading to

$$g = \frac{Gm_E}{r_E^2}.$$

Gravitational field strength,

$$g = \frac{\text{force acting on small test object}}{\text{mass of test object}}$$

$$= \frac{F}{m}.$$

The symbol g here does not refer specifically to the Earth's surface but anywhere in a gravitational field.

Two units can be used for g: N kg^{-1} (when dealing with gravitational field strength) and m s^{-2} (when dealing with acceleration).

Test objects are required because the object itself contributes to the gravitational field where it is placed. Their magnitude must be much smaller than that of the mass producing the field. This is usual when defining field strengths.

Much of the theory of satellite motion is still correct when a satellite has an elliptical orbit (the usual case). The assumption in DP physics is that satellites have circular orbits.

Equating centripetal gravitational forces (using $v = r\omega$ and $\omega = \dfrac{2\pi}{T}$) gives

$$m\omega^2 r = \frac{4\pi^2 mr}{T^2} = \frac{mv^2}{r} = \frac{GMm}{r^2},$$

where M is the planet mass, m is the satellite mass, r is the orbital radius that separates M and m, v is the linear orbital speed, ω is the orbital angular speed and T is the orbital period.

Also, $v = \sqrt{\dfrac{GM}{r}}$, $T = 2\pi\sqrt{\dfrac{r^3}{GM}}$

and $\omega = \sqrt{\dfrac{GM}{r^3}}$.

You should be able to derive these equations and use them.

Solution

a) The satellite has a periodic time of 24 hours $\equiv 86\,400$ s.

T for the orbit is $\dfrac{2\pi}{\omega}$. So the angular speed

$$\omega = \frac{2\pi}{T} = 7.3 \times 10^{-5} \text{ rad s}^{-1}.$$

b) $\omega = \sqrt{\dfrac{GM}{r^3}}$, $r = \sqrt[3]{\dfrac{GM}{\omega^2}} = \left(\dfrac{6.7 \times 10^{-11} \times 6.0 \times 10^{24}}{\left(7.3 \times 10^{-5}\right)^2}\right)^{\frac{1}{3}} = 42$ Mm.

SAMPLE STUDENT ANSWER

The two arrows in the diagram show the gravitational field strength vectors at the position of a planet due to each of two stars of equal mass M.

planet

6.8 × 10¹¹ m

star   star

Each star has mass $M = 2.0 \times 10^{30}$ kg. The planet is at a distance of 6.0×10^{11} m from each star.

a) Show that the gravitational field strength at the position of the planet due to one of the stars is $g = 3.7 \times 10^{-4}$ N kg⁻¹.

This answer could have achieved 1/1 marks: [1]

▲ The equation is quoted and the substitution is clear. The answer has an appropriate number of significant figures (one more than the 2 sf in the question) and there is a correct unit for the answer.

$$g = G\frac{m}{r^2} = \frac{(6.67 \times 10^{-11}) \times (2 \times 10^{30})}{\left(6.0 \times 10^{11}\right)^2} = 3.71 \times 10^{-4} \text{ N kg}^{-1}$$

b) Draw two arrows to show the gravitational field strength at the position of the planet due to each of the two stars.

This answer could have achieved 1/2 marks: [2]

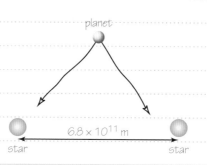

planet

6.8 × 10¹¹ m

star   star

▼ The arrows are drawn poorly; the right-hand arrow clearly misses the edge of the star. Greater care is needed with diagrams to score full marks.

c) Calculate the magnitude and state the direction of the resultant gravitational field strength at the position of the planet.

This answer could have achieved 1/3 marks: [3]

▼ The value in part a) has to be combined, having calculated the angle between the stars and dividing it by two (calculating the gravitational field strength again is not needed). However, 4.94×10^{11} is not the value quoted.

$$F = \frac{GM}{r^2}$$

$G = 6.67 \times 10^{-11}$ Nm²/kg² $M = 4.0 \times 10^{30}$ $r = 4.94 \times 10^{11}$

$$F = \frac{2.67 \times 10^{20} \text{ Nm}^2}{2.44 \times 10^{23} \text{ m}}$$

$F = 4.48 \times 10^{-27}$ Newtons towards the barycenter of the two stars.

▲ The direction is correct: 'barycentre' is another word for centre of mass of two (or more) bodies and is midway between the stars.

Practice problems for Topic 6

Problem 1

A toy train moves with a constant speed on a horizontal circular track of constant radius.

a) State and explain the direction of the horizontal force that acts on the train.

b) The mass of the train is 0.24 kg. It travels at a speed of 0.19 m s^{-1}. The radius of the track is 1.7 m.

Calculate the centripetal force acting on the engine.

Problem 2

Two charged objects, X and Y, are positioned so that the gravitational force between them is equal and opposite to the electric force between them.

a) State and explain what happens when the distance between the centres of X and Y is doubled.

b) The mass of X is now doubled with no other changes. Deduce how the charge on Y must change so that the resultant force is unaltered.

Problem 3

An Earth satellite moves in a circular orbit.

Explain why its speed is constant, even though a force acts on it.

Problem 4

A beta particle is emitted from a nucleus, P. The subsequent path of the beta particle is part of a circle of radius 0.045 m.

The speed of the electron is 4.2×10^7 m s^{-1}.

a) Calculate the momentum of the electron.

b) Calculate the magnitude of the force acting on the electron that makes it follow the curved path.

c) Identify the direction of this force.

7.1 DISCRETE ENERGY AND RADIOACTIVITY

You must know:

✔ what is meant by discrete energy and discrete energy levels

✔ what is meant by a transition between two energy levels

✔ that radioactive decay is a random, spontaneous process during which radioactive emissions occur and the nucleus of an atom changes

✔ that there are four fundamental forces: electromagnetic, gravitational, strong nuclear and weak nuclear

✔ the properties of alpha and beta particles and gamma radiation and the decay equations that describe the changes to the nucleus

✔ what is meant by background radiation.

You should be able to:

✔ describe the emission and absorption spectrum of common gases

✔ solve problems involving atomic spectra, including calculations of the wavelength of photons emitted during atomic transitions

✔ complete decay equations for alpha and beta decay and describe the absorption characteristics of decay particles

✔ describe what is meant by half life and determine the half-life of a nuclide from a decay curve

✔ investigate half-life experimentally (or by a simulation)

✔ demonstrate that isotopes of an element have the same number of protons but different numbers of neutrons.

A neutral *atom* consists of a positively-charged *nucleus* with *electrons* outside it. The nucleus contains positively charged *protons* and neutral *neutrons*. In neutral atoms, the number of protons equals the number of electrons. An *ion* forms when a neutral atom gains or loses one or more electrons.

The energy levels of an atom are *discrete*, having a lowest *ground state*. The levels lead to *emission spectra* that are also discrete and arise when energy is transferred to the atom and is subsequently re-emitted.

🔗 The Planck constant is covered in more detail in Topic 12.

The **energy of a photon** E is related to its frequency f and wavelength λ

by $E = hf = \dfrac{hc}{\lambda}$ or $\lambda = \dfrac{hc}{E}$

where c is the speed of light and h is the Planck constant.

Example 7.1.1

The energy levels of the hydrogen atom are shown.

a) Determine, in eV, the energy of a photon of wavelength 658 nm.

b) Identify the transition that gives rise to a photon of this wavelength.

c) Explain why the lines in an emission spectrum involving only the ground state become closer together as the wavelength of the emitted photons decreases.

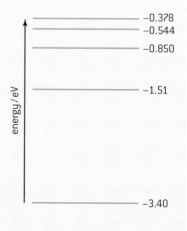

energy / eV

−0.378
−0.544
−0.850
−1.51
−3.40

Solution

a) $E = \dfrac{hc}{\lambda} = \dfrac{6.63 \times 10^{-34} \times 3.00 \times 10^8}{658 \times 10^{-9}} = 3.02 \times 10^{-19}\,\text{J}$

To convert to eV:

$\dfrac{3.02 \times 10^{-19}}{1.60 \times 10^{-19}} = 1.89\,\text{eV}$

b) The 1.89 eV must be the energy difference between the two levels concerned. The only possibility on the diagram is the change between −1.51 and −3.40, which are the ground state and the first excited state.

c) At higher energy levels, the energy levels become closer together. The energy differences between the first excited level and the second excited level and above become more equal and the difference in the wavelength of the emitted photons decreases too.

> **Isotopes** of an element have the same number of protons but a different number of neutrons. Isotopes have the same chemistry but different physics.

Many elements have *isotopes* which undergo natural *radioactive decay*. During radioactive decay *ionizing radiation* is emitted and the nature of the nucleus changes.

Radioactive decay is random and spontaneous. *Random* means that we cannot predict when a nucleus will decay; *spontaneous* means that radioactive decay cannot be affected by changing the temperature, pressure, and so on.

Some of the emissions in radioactive decay, together with some of their properties, are listed in this table:

>> Assessment tip

Radioactive decay and nuclear fission can be confused. Decay occurs naturally; most fission events are induced by the arrival of a neutron (Topic 7.2).

	Alpha α	Beta β^-	Beta β^+	Gamma γ
Nature	Helium nucleus; 2p + 2n	Electron from the nucleus; β^-; e^-	Positron (anti-electron) from the nucleus; β^+; e^+	Electromagnetic radiation
Origin	Overall change in proton: neutron ratio	Change of d quark to u quark as neutron changes to proton	Change of u quark to d quark as neutron changes to proton	Removal of excess energy from nucleus
Ionizing power	Strong	Medium		Weak
Penetration	Few cm of gas; few mm of paper	Few cm of aluminium	Quickly annihilated with electron	Many metres of lead or concrete
Notes	One or two energy states for a decay	Broad energy spectrum; an electron antineutrino is emitted in addition	Broad energy spectrum; an electron neutrino is emitted in addition	Discrete energies tied to nuclear energy levels
Decay equation	${}^{A}_{Z}X \rightarrow {}^{A-4}_{Z-2}Y + {}^{4}_{2}\alpha$	${}^{A}_{Z}X \rightarrow {}^{A}_{Z+1}Y + {}^{0}_{-1}\beta^- + \overline{\upsilon_e}$	${}^{A}_{Z}X \rightarrow {}^{A}_{Z-1}Y + {}^{0}_{+1}\beta^+ + \upsilon_e$	${}^{A}_{Z}X^* \rightarrow {}^{A}_{Z}X + \text{energy}$

The notation used in the table is: ${}^{A}_{Z}X$, where X is the chemical symbol for the element, A is the nucleon number (total number of protons + neutrons in nucleus), Z is the proton number (total number of protons in nucleus). Sometimes the symbol N is used for the neutron number.

$A = Z + N$

> **>> Assessment tip**
>
> Equations for decay processes must balance both for A and Z. Use the correct notation for the neutrino and the antineutrino.

> Some of the table entries are covered in later topics (7.3, 12.1 and 12.2).

>> **Assessment tip**

You should be able to list the sources of background radiation. The count rates from radioactive sources used in laboratory experiments need to be corrected for the background. Example 7.1.2 shows how this is done.

Half-life is the time taken for half of the atoms initially present in a pure sample of a radioactive nuclide to decay. It can be determined from a graph of *corrected count rate* against *time*.

>> **Assessment tip**

Half-life is a value that relates to a large sample of decaying atoms. An individual nucleus does not have half-life. Topic 12.1 introduces the idea of a decay constant which is the probability that an individual nucleus will decay in the next second.

Activity is the total rate at which a sample is decaying.

The unit of **activity** is the **Becquerel** (abbreviated Bq). 1 Bq is the activity of one disintegration every second.

Count rate is the measured number of counts being detected in one second.

Activity and count rate are different; it is difficult to count every emission from a decaying sample. Some emissions will be absorbed by the sample itself.

Natural sources of radiation give rise to *background radiation*.

Each nucleus of a naturally decaying element has an identical chance of decay per second. The total number of decays in one second, the *decay rate,* is proportional to the number of nuclei of the element in a sample. This leads to behaviour where the time to halve the number of original nuclei is constant, irrespective of how many were originally in the sample. This time is known as the *half-life*.

Example 7.1.2

The decay activity of a pure radioactive sample was measured every 30 s. The background activity is 0.4 Bq.

Time t / s	0	30	60	90	120
Count rate / Bq	14.0	9.6	6.6	4.6	3.2

Determine the half-life of the sample.

Solution
Subtract the background activity (0.4 Bq) from each count rate value.

Time, t / s	0	30	60	90	120
Count rate / Bq	14.0	9.6	6.6	4.6	3.2
Count rate minus background activity	13.6	9.2	6.2	4.2	2.8

Plot these values on a graph:

Determine the half-life for at least three different time intervals on the graph (this graph shows the half-lives from 14 to 7, 10 to 5 and 6 to 3 counts per second).

The (mean) average of these three values is 52 s.

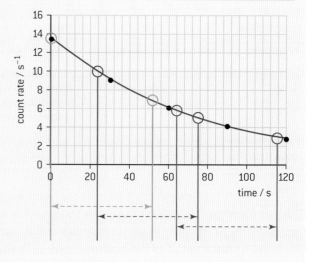

Nuclear properties are governed by the fundamental forces that act between particles in the nucleus. In order of increasing strength, where 1 is the strongest force, these are:

There are examples of questions using nuclear forces in Topic 7.3.

Gravity	• acts between masses
	• operates to infinity
	• weakest known force; relative strength 10^{-38}
Weak nuclear force	• acts on quarks and leptons
	• governs decay of nucleons
	• force carriers are W and Z particles
	• operates within nucleus only (short range)
	• relative strength 10^{-6}
Electromagnetic force	• acts between charged particles
	• force carrier is the photon
	• operates to infinity
	• relative strength 10^{-2}
Strong nuclear force	• acts between quarks and gluons
	• force carrier is gluon
	• operates within short range
	• strongest known force; relative strength 1.

SAMPLE STUDENT ANSWER

The graph shows the variation with time t of the activity A of a sample containing phosphorus-32 $\left(^{32}_{15}P\right)$.

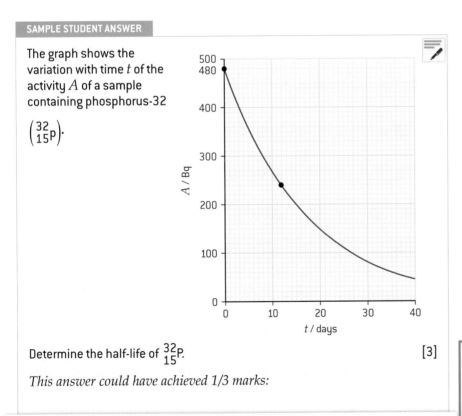

Determine the half-life of $^{32}_{15}P$. [3]

This answer could have achieved 1/3 marks:

12 days

▲ The answer is correct.

▼ This is a 'determine' question. The command verb implies that full details are required for the answer. A close look at the graph shows that there are markings at a count rate of 480 Bq and at 240 Bq. So this answer probably comes from only one determination. This view is supported by the lack of any quoted averages.

7.2   NUCLEAR REACTIONS

You must know:

✔ the definition of the unified atomic mass unit

✔ that mass defect and nuclear binding energy are ways to describe energy stored in the nucleus

✔ what is meant by nuclear fission and by nuclear fusion

✔ that the equation $\Delta E = c^2 \Delta m$ can be applied to nuclear changes.

You should be able to:

✔ solve problems involving mass defect and binding energy

✔ solve problems involving the energy released in radioactive decay, nuclear fusion and nuclear fission

✔ sketch and interpret the general shape of the curve of average binding energy per nucleon against nucleon number.

The magnitude of the **unified atomic mass unit** (abbreviated u) is very close to the mass of a proton or neutron (these differ slightly from each other).

1 u is defined to be the mass of $\dfrac{1}{12}$ of the mass of a stationary carbon-12 atom.

Energy and mass are considered interchangeable using $\Delta E = c^2 \Delta m$: mass can be expressed in energy units: a kilogramme is equivalent to $(3 \times 10^8)^2$ or 9×10^{16} J.

$1\,\text{eV} \equiv 1.6 \times 10^{-19}\,\text{J}$

Therefore, one electron volt is equivalent to 1.8×10^{-36} kg.

When expressed using eV in this way, the unit of mass is $\text{eV}\,c^{-2}$.

So $2\,\text{eV}\,c^{-2} \equiv 3.6 \times 10^{-36}$ kg.

🔗 The fission reaction can self-sustain and is used in nuclear power stations. Details of these stations are given in Topic 8.1.

The *unified atomic mass unit* is used in nuclear physics. Mass can be expressed in energy terms.

Few elements undergo *spontaneous nuclear fission*. More commonly, fission is **induced** when a moving neutron interacts with a nucleus.

Nuclear fission occurs for a uranium (U) nuclide. A neutron collides with a nucleus of U-235 and is absorbed. This produces a nucleus of the highly unstable U-236. The U-236 splits into two (or more) nuclear fragments with the release of several fast-moving neutrons.

Fusion also leads to energy release. This is the origin of the energy in a star, such as the Sun. The stellar hydrogen is at such a high temperature that electrons are stripped away from atoms to leave single protons and electrons in a *plasma*.

Figure 7.2.1 shows the cycle in which two protons fuse to produce a deuterium (H-2) nucleus (an isotope of hydrogen) with the release of a positron (an anti-electron) and an electron neutrino. Further fusions with protons occur forming first helium-3 (He-3) and, finally, He-4. There is an overall energy release during the cycle.

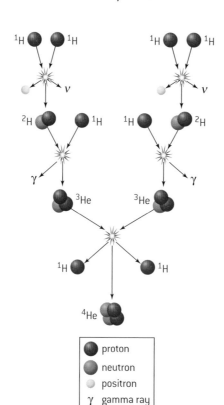

●	proton
●	neutron
○	positron
γ	gamma ray
ν	neutrino

▲ **Figure 7.2.1.** The fusion reaction that produces helium from hydrogen

Example 7.2.1

Uranium-238 undergoes alpha decay to form thorium-234. The table shows the masses of the uranium-238 $\left(^{238}_{92}U\right)$ and thorium-234 nuclei and of an alpha particle (helium-4).

The following data are available:

Mass of uranium-238 = 238.0002 u

Thorium-234 = 233.9941 u

Alpha particle = 4.0015 u

a) i) State the number of neutrons in a uranium-238 nucleus.

   ii) State the number of protons in a nucleus of thorium.

b) i) Determine the mass change, in kg, when a nucleus of uranium-238 undergoes alpha decay.

   ii) Determine the increase in kinetic energy of the products in part b) i).

Solution

a) i) 238 − 92 = 146

   ii) The helium nucleus is ^4_2He, so two of the protons from $^{238}_{92}U$ are removed when the thorium forms.

   There are, therefore, 90 protons.

b) i) The mass change is 238.0002 − 233.9941 − 4.0015 = 0.0046 u.

   $0.0046 \times 1.661 \times 10^{-27} = 7.6 \times 10^{-30}$ kg

   ii) This can be done via $\Delta E = c^2 \Delta m$ or by remembering that 1 u ≡ 931.5 MeV.

   $0.0046 \times 931.5 = 4.3$ MeV.

> **binding energy per nucleon**
>
> $$= \frac{\text{total binding energy for a nucleus}}{\text{number of protons + number of neutrons}}$$

Precise determinations of atomic mass show that the mass of a nucleus is less than the total mass of its constituent parts (proton mass plus neutron mass). This difference is known as the *mass defect*. This can also be described in energy terms through the equivalence of mass and energy and is then known as the *binding energy*.

Binding energy has its origins in the forces inside the nucleus. Protons are positively charged and repel each other electrostatically. However, a *strong nuclear force* acts between the quarks which make up protons and neutrons. At small distances, the strong force is repulsive but, at larger distances within the nucleus, it is attractive and binds the protons and neutrons together.

Energy must be added to a nucleus to separate it into its component parts. Similarly, to form a nucleus by bringing its individual nucleons together from infinity requires the removal of energy. This is the source of the binding energy and mass defect.

Plotting a graph of **binding energy per nucleon** against **nucleon number** demonstrates nuclear stability.

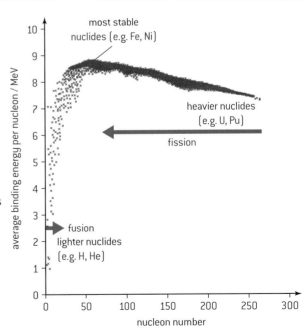

▲ **Figure 7.2.2.** Graph of **binding energy per nucleon** against **nucleon number**

When a large nucleus, such as U-236, undergoes fission, the two smaller fragments are near the highest part of the curve. These fragments have a larger magnitude of binding energy per nucleon (they are more strongly

The **change** in binding energy **per nucleon** for each reaction is greater for fusion than fission. On the face of it, fusion will be better than fission for power generation. Unfortunately, many of the engineering problems raised by nuclear fusion have yet to be solved.

bound) and therefore this energy must be transferred out of the system. The system moves to a more stable state.

In fusion, hydrogen nuclei (zero binding energy per nucleon) form a helium nucleus (more stable; larger binding energy per nucleon). Again, the nuclei lose energy.

SAMPLE STUDENT ANSWER

A nucleus of phosphorus-32 $\left(^{32}_{15}P\right)$ decays by beta minus (β⁻) decay into a nucleus of sulphur-32 $\left(^{32}_{16}S\right)$. The binding energy per nucleon of $^{32}_{15}P$ is 8.398 MeV and for $^{32}_{16}S$ it is 8.450 MeV.

Determine the energy released in this decay.   [2]

This answer could have achieved 1/2 marks:

$32 \times 8.398 = 268.74$

$32 \times 8.45 = 270.4$

$M_e = 0.511 \, \text{MeVC}^{-2}$

$\quad = 2.171 \, \text{MeV}$

▼ The mistake here is to include the mass of the beta-minus particle in the calculation. The particle is not bound to anything else so has no binding energy. In the same way, hydrogen-1 has zero binding energy as it consists of one proton in the nucleus.

This answer could have achieved 2/2 marks:

$(8.450 \times 32) - (8.398) \times (32) = 32(8.450 - 8.398)$

$= 1.664 \, \text{MeV}$

▲ Although not at all well explained, this does arrive at the correct answer and scores full marks even though this was a "determine" question, where a high quality of explanation is expected.

7.3   THE STRUCTURE OF MATTER

You must know:

✔ quarks are used to explain the patterns in nuclear particles

✔ leptons have two forms – charged and uncharged – and that neither form takes part in strong interactions

✔ the Standard Model uses six quarks and six leptons to describe our present view of matter

✔ there are two families of hadrons: baryons with three quarks, and mesons that are a quark–antiquark pair

✔ the conservation laws of charge, baryon number, lepton number and strangeness

✔ the nature and range of the strong nuclear, weak nuclear and electromagnetic forces

✔ there are exchange particles and how they mediate the fundamental forces.

You should be able to:

✔ use the conservation of charge, baryon number, lepton number and strangeness to solve problems involving particle interactions

✔ describe protons and neutrons in terms of quarks

✔ compare the interaction strengths of the fundamental forces including gravity

✔ describe the mediation of the fundamental particles through exchange particles

✔ sketch and interpret Feynman diagrams

✔ describe why free quarks are not observed

✔ describe the details of the Rutherford–Geiger–Marsden experiment that led to the discovery of the nucleus

✔ what constitutes a Feynman diagram

✔ describe confinement and explain why free quarks are not observed

✔ explain that there is a Higgs boson which accounts for the mass of quarks and charged leptons.

In the early 20th century, Rutherford suggested that Geiger and Marsden should investigate alpha particle scattering. They found that a small number of the alpha particles were scattered through very large angles by a thin gold foil. This result allowed Rutherford to propose that the nucleus is small, dense and positively-charged with the atomic electrons outside the nucleus.

Later in the century, it became clear that there were both positive (protons) and neutral (neutron) particles in the nucleus. Experiments then began to reveal an increasing complexity in the interactions in the nucleus until, in the mid-1970s, the Standard Model was agreed and confirmed by particle physicists.

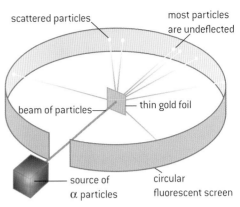

▲ **Figure 7.3.1.** The Rutherford–Geiger–Marsden experiment

Example 7.3.1

In the Geiger and Marsden experiment, most alpha particles pass through gold foil with no change in direction or energy loss. A small number of particles deviate from their original direction by angles more than 90°.

Explain, with reference to Rutherford's atomic model and the forces acting in the nucleus:

a) why some alpha particles are deflected through large angles

b) why most of the alpha particles are not deviated.

Solution

a) A few alpha particles approach very close to a gold nucleus. Both the nucleus and the alpha particle are positively charged and so there is a repulsive force between them.

The force varies as $\dfrac{1}{r^2}$, where r is the distance between the centres of the alpha particle and the gold nucleus.

When r is very small, the force is very large and the alpha particle undergoes a significant deviation.

b) According to the Rutherford model, the space between the gold nuclei is very large compared with their diameter. The majority of the alpha particles do not approach a gold nucleus closely enough to experience a significant force.

> The Standard Model uses three groups of particles to describe interactions within nucleons.
>
> **Quarks:** These combine to form **hadrons** which divide into the sub-groups **baryons** (which contain three quarks) and **mesons** (which contain two quarks: a quark–antiquark pair).
>
> **Leptons:** These cannot combine with each other.
>
> **Exchange particles:** these convey information from quark to lepton and between members of each group.

Quarks and *leptons* are classified by their electric charge and rest mass. They are grouped in 'generations' with mass increasing with generation number. Every particle has an antiparticle with opposite charge (neutrinos are uncharged but still have antiparticles).

This table gives the names and other details of the particles.

	Charge	First generation	Second generation	Third generation	Baryon number	Lepton number
Quarks	$+\dfrac{2}{3}e$	up (u)	charm (c)	top (t)	$\dfrac{1}{3}$	
	$-\dfrac{1}{3}e$	down (d)	strange (s)	bottom (b)	$\dfrac{1}{3}$	
Leptons	0	electron neutrino (v_e)	muon neutrino (v_μ)	tau neutrino (v_τ)		+1 for leptons
	$-e$	electron (e⁻)	negative muon (μ^-)	tau (τ)		−1 for antileptons

→ *increasing mass*

Baryons	proton	p	uud
	neutron	n	udd
Mesons	neutral kaon	K^0	$d\bar{s}$
		$\overline{K^0}$	$s\bar{d}$
	positive kaon	K^+	$u\bar{s}$
	negative kaon	K^-	$s\bar{u}$
	neutral pion	π^0	$u\bar{u}$ or $d\bar{d}$
	positive pion	π^+	$u\bar{d}$
	negative pion	π^-	$d\bar{u}$

The conservation rules for particle interactions involve:

• charge

• baryon number

• lepton number

• strangeness.

These must balance on both sides of the equation for a particle interaction to be possible.

Quarks have strangeness number = 0, except for the strange quark that has strangeness number –1.

The constitution of some baryons and mesons is given in this table.

Conservation laws allow us to decide whether a proposed particle reaction is possible or not.

Example 7.3.2

Reaction ① $p + n \rightarrow p + n + p + \bar{p}$

Reaction ② $p + n \rightarrow p + \mu^+ + \mu^-$

Explain, in terms of baryon conservation, for the interactions between a proton (p) and a neutron (n), why:

a) reaction ① is possible

b) reaction ② is impossible.

Solution

a) The baryon number on the left-hand side is $1 + 1 = 2$.

The right-hand side is $1 + 1 + 1 - 1 = 2$ and so this is possible.

b) The baryon number for a meson is $+\dfrac{1}{3} - \dfrac{1}{3} = 0$.

The baryon number on the left-hand side is $1 + 1 = 2$ whereas the baryon number on the right is 1 so this would violate conservation of baryon number.

Confinement (also known as colour confinement) means that free quarks are never observed. This is because quarks are subject to the strong interaction. As energy is supplied to a two-quark system, the quarks separate but the force acting between them does not decrease. Contrast this behaviour with that of the electromagnetic field where force is proportional to $\dfrac{1}{r^2}$. For quarks, there comes a point where it is energetically more favourable to create a new quark–antiquark pair rather than to completely separate and, therefore, observe individual quarks.

The interpretation of this effect in the Standard Model is via the quark flavour of 'colour'. Each flavour can have three colours, described as red, blue and green. The antiquarks have anti-red, anti-blue and anti-green colours. Baryons consist of three quarks making white; mesons must consist of a colour–anti-colour pair. Thus all hadrons are colourless. Quarks are always observed bound in colourless groups when they emerge in large-energy collisions.

As quarks move closer together under conditions of high energy, the force between them becomes small. This is known as *asymptotic freedom*.

The existence of the *Higgs boson* was postulated in the 1960s.

In the Higgs theory, mass is a property conferred on particles when they interact with the Higgs field.

One way to explain the Higgs field is to use an analogy with the refractive index of a transparent medium. We know the medium is there because different wavelengths travel through it at different speeds. We could say that blue photons are heavier than red!

Similarly, different particles interact with the Higgs field to different degrees. We can think of the quark mass as being due to a 'dynamic friction' between it and the Higgs field.

Feynman diagrams are combinations of vertices. A vertex has at least one arrow pointing to it and at least one away from it. The vertex represents either a quark–quark transition or a lepton–lepton transition. The arrowhead direction is crucial. The diagrams usually have time going from left to right (some diagrams are drawn with time going bottom to top).

Photons and W and Z particles are shown as *wavy lines*.

Quarks or leptons are shown as *straight lines*.

antiquark
(or antilepton)
entering the vertex

vertex

exchange particle

quark
(or lepton)
entering the vertex

time

▲ **Figure 7.3.2.** A vertex in a Feynman diagram

Example 7.3.3

In electron capture (inverse beta-minus decay), a proton and an electron interact to form a neutron and another particle.

Draw the Feynman diagram to represent this interaction.
Identify the particles involved.

Solution

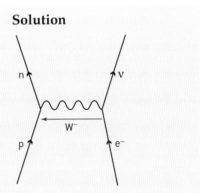

Feynman diagrams have mathematical significance for particle physicists. In the DP physics course, they are used to illustrate the interactions between particles and exchange particles.

a) A particular K meson has a quark structure ūs. State the charge on this meson. [1]

This answer could have achieved 0/1 marks:

▼ The answer is that the ū has a charge of $-\frac{2}{3}e$ and the s $-\frac{1}{3}e$. The overall charge is –*e*. Just writing "negative" is not enough to score.

negative

b) The Feynman diagram shows the changes that occur during beta minus (β⁻) decay.

Label the diagram by inserting the **four** missing particle symbols. [2]

This answer could have achieved 1/2 marks:

▲ The change of a d quark to a u quark is correct.

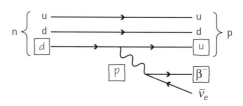

▼ The remaining boxes include an emitted proton. This is not correct in a Feynman diagram even though on a baryon level the proton is given out.

c) Carbon-14 (C-14) is a radioactive isotope which undergoes beta minus (β⁻) decay to the stable isotope nitrogen-14 (N-14). Energy is released during this decay. Explain why the mass of a C-14 nucleus and the mass of a N-14 nucleus are slightly different even though they have the same nucleon number. [2]

This answer could have achieved 0/2 marks:

▼ The energy released must have come from the transfer of mass to energy. The nitrogen nucleus is more stable and therefore has more binding energy and a smaller mass. This is not just the result of particle exchange. The answer does not make this clear and focuses on the difference in mass of a proton and a neutron.

During the beta decay, a neutron is turned into a proton, which means the product (N-14) has one more proton than Carbon-14 which causes their mass to be slightly different. However the nucleon number (neutrons+protons) stays the same, because for a beta decay, a neutron turns into a proton.

Practice problems for Topic 7

Problem 1

A student determined the half-life of a radioactive nuclide by placing it near a detector. He recorded the number of counts in 30 seconds every 10 minutes from the start of the experiment.

The results given in the table were obtained.

t / minute	0	10	20	30	40	50	60
Number of counts in 30 s	60	42	35	23	18	14	10

a) Explain what is meant by *half-life*.

b) Determine the half-life for this nuclide.

Problem 2

The energy levels of an atom are shown.

0 _____

-2.42×10^{-19} J _____ level 2

-5.48×10^{-19} J _____ level 1

-2.18×10^{-18} J _____ ground state

An electron with a kinetic energy of 2.0×10^{-18} J makes an inelastic collision with an atom in the ground state.

a) Calculate the speed of the electron just before the collision.

b) **(i)** Deduce whether the electron can excite the atom to level 2.

(ii) Calculate the wavelength of the radiation that will result when an atom in level 2 falls to level 1.

(iii) State the region of the spectrum in which the radiation in part b) ii) belongs.

Problem 3

Plutonium-240 $\left(^{240}_{94}\text{Pu}\right)$ decays to form uranium (U) and an alpha-particle (α).

The following data are available:

Mass of plutonium nucleus $= 3.98626 \times 10^{-25}$ kg

Mass of uranium nucleus $= 3.91970 \times 10^{-25}$ kg

Mass of alpha particle $= 6.64251 \times 10^{-27}$ kg

Speed of electromagnetic radiation $= 2.99792 \times 10^8 \text{ m s}^{-1}$

a) State the equation for this decay.

$$^{240}_{94}\text{Pu} \rightarrow$$

b) Determine the energy released when one nucleus decays.

Problem 4

a) **(i)** Identify, using the graph of binding energy per nucleon versus nucleon number (Figure 7.2.2), the nucleon number and the nuclear binding energy per nucleon for the most stable nuclide.

(ii) Calculate the binding energy of the nuclide in part a) (i).

b) Two protons fuse to form a deuterium nucleus as described by:

$$^1_1\text{H} + ^1_1\text{H} \rightarrow ^2_1\text{H} + ^0_1\text{e} + \nu + 1.44 \text{ MeV}$$

Identify:

(i) ^0_1e

(ii) ν

(iii) Show that this reaction satisfies the conservation laws for charge, baryon number and lepton number.

Problem 5

State the quark composition of

a) a proton

b) a positive pion, π^+.

Problem 6

State, with a reason, whether the following particle reactions are possible.

a) $p + \pi^- \rightarrow K^- + \pi^+$

b) $p + \bar{\nu} \rightarrow n + e^+$

8 ENERGY PRODUCTION

8.1 ENERGY SOURCES

You must know:

✔ the meaning of specific energy and energy density for energy sources

✔ what a Sankey diagram is

✔ the distinction between primary and secondary energy sources

✔ electricity is a versatile form of secondary energy

✔ renewable and non-renewable energy sources

✔ a thermal nuclear reactor requires a moderator, a method of control and a way to exchange thermal energy.

You should be able to:

✔ solve specific energy and energy density problems

✔ sketch and interpret Sankey diagrams

✔ describe the basic features of, and the energy transformations in, power stations based on the transfer of energy from fossil fuels, nuclear fuels, the wind, solar energy and water-based systems

✔ discuss safety issues and risks associated with the production of nuclear power

✔ describe the differences between solar power cells and solar photovoltaic cells.

Fuels can be characterised as renewable or non-renewable. Fossil fuels regenerate in timescales of millions of years and cannot be replenished as quickly as they are consumed.

Some sources can be regenerated rapidly; wind, wave, solar energy and biomass are examples. One possible definition of a renewable resource is that it can be regenerated at the same rate as that at which it is used up.

Much of the energy we use originates in the Sun. This includes:

- fossil fuels that were once produced via photosynthesis
- the direct transfer of energy from radiation arriving at the Earth's surface
- the generation of weather and tidal systems that are harnessed for energy transfer.

These are known as *primary energy sources*. Other primary sources include nuclear fuel and geothermal energy.

Other types of energy source involve an intermediate transfer step before the energy can be used; these are *secondary sources*. They include electrical energy and some chemicals, such as petrol, where refinement of the crude fossil fuel is required before use.

Some energy sources are *renewable*. Others cannot be regenerated within a realistic time and are said to be *non-renewable*.

Example 8.1.1

A coal-fired power station has a power output P. Its efficiency is ε. It burns a mass of coal M of density ρ every second. Derive an expression for the energy density of the coal.

Solution

Energy density is the energy available from unit volume of the fuel.

Use $\rho = \dfrac{M}{V}$

The volume V of coal consumed every second is $\dfrac{M}{\rho}$.

The energy input to the station is $\dfrac{P}{E}$, allowing for the inefficiency.

The energy density is $\dfrac{\text{energy input}}{\text{coal volume}} = \dfrac{P}{E} \times \dfrac{\rho}{M} = \dfrac{P\rho}{\varepsilon M}$.

> **Specific energy** is the energy that can be transferred from 1 kg of the fuel (unit: $\mathrm{J\,kg^{-1}}$).
>
> **Energy density** is the energy that can be transferred from 1 m³ of the fuel (unit: $\mathrm{J\,m^{-3}}$).

Thermal power generating stations use thermal energy to boil water, creating steam, which turns a turbine—this is a transfer of thermal energy to mechanical (kinetic) energy. The turbine is attached to a dynamo which rotates in a magnetic field. This causes charge to flow in the dynamo's coil, transferring the kinetic energy to an electrical form. The electrical energy can be transported through a cable network (a grid) to domestic and commercial end-users.

The initial thermal energy can be transferred from various fuels, including nuclear, geothermal and fossil fuels.

One way to represent the quantitative energy transfers is to use a *Sankey diagram*.

> **Sankey diagrams** are a type of flow diagram that show energy transfers in a system or process. The width of an arrow in the diagram represents the relative size of its contribution to the total energy involved in the transfer.

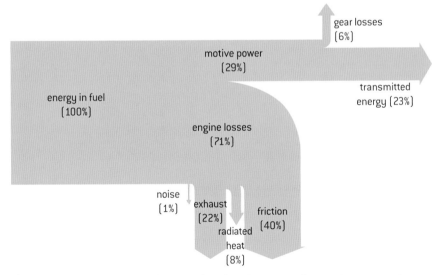

▲ **Figure 8.1.1.** A Sankey diagram showing the energy losses in a car engine

Nuclear power stations use nuclides that undergo induced fission.

Neutrons bombard the uranium nuclei leading to the formation of smaller, highly radioactive nuclei and more neutrons. The emitted neutrons initiate further fissions, leading to a chain reaction. The reaction is controlled using *control rods* made from a material that absorbs neutrons.

The neutrons released during fission are emitted at high speeds. But neutrons with low speeds have the highest probability of causing further fissions. So emitted neutrons are slowed down in a *moderator*. Moderator atoms have nuclei with which neutrons collide elastically.

> The physics of nuclear fission is described in Topic 7.2.

The thermal energy released during fission is used to boil water for the turbine steam.

In *fossil-fuel stations*, the fuel is burned directly releasing thermal energy to boil water for the turbine steam.

In *geothermal stations*, water at high pressure is forced below the Earth's surface to regions where the rocks are at high temperatures.

The water is heated and returned to the surface as steam for a turbine. The water is then reused in the cycle. The hot water can also be used in a heat exchanger to supply thermal energy to nearby premises.

The technique is mainly used where local geology has created high temperatures close to the surface.

Solar power uses two methods to transfer energy from solar radiation.

Solar heating panels contain circulating water (usually with an antifreeze agent added) that transfers energy from the hot panel in the sunlight to a water storage tank. The energy stored in the water can then be used via a heat exchanger to produce hot water for domestic use.

Solar photovoltaic cells use semiconductor materials to transfer photon energy from sunlight to an electrical form. This generates an alternating current to use locally or to supply the grid system.

Take care with the distinction between solar heating panels and solar photovoltaic cells. The former do not have an electrical energy output; they store thermal energy in water. The photovoltaic cells are the type that transfer the energy arriving with the photons into an electrical form.

> The physics of the energy transfer from generating station to user is discussed in Topic 11.2.

> The flow equation does not represent the maximum power that can be *extracted* from the turbine. The equation assumes a final speed of zero for the fluid and the fluid must have some kinetic energy remaining to move it away from the turbine. There have been several estimates for this residual energy; one of the earliest is due to Albert Betz. He suggested that no turbine can transfer more than $\dfrac{16}{27}$ (about 60%) of the theoretical maximum power.

Example 8.1.2

Outline why most of the world's energy consumption is provided by fossil fuels.

Solution
Fossil fuels are widely available and large quantities still remain in the ground. They can be transported easily or power stations can be built close to the fuel source. Fossil fuels have a high energy density, so relatively small volumes produce significant amounts of energy.

Fluid flow through a turbine is used to generate renewable energy. It can be used with either **wind** or **water**. The same *flow equation* is used for both. When the speed of the fluid is v and the density of the fluid is ρ, the power available from the fluid arriving at a turbine of cross-sectional blade area A is $\dfrac{1}{2}A\rho v^3$.

Wind turbines can be in a horizontal or vertical format (see Figure 8.1.2).

Water turbines can generate electrical energy in several ways. These include:

- hydroelectric systems

- pumped storage hydroelectric systems

- tidal techniques (using the movement of water bodies due to Moon and Sun tides)

- wave techniques (using the movement of waves onto a shore).

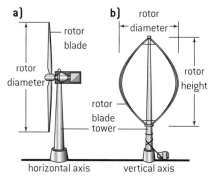

▲ **Figure 8.1.2.** Horizontal and vertical wind turbines

Example 8.1.3

A wind turbine's blades have a total area $65\,\text{m}^2$.

The turbine is used in a wind of speed $15\,\text{m s}^{-1}$.

The density of air is $1.3\,\text{kg m}^{-3}$.

a) Determine the mass of air incident on the turbine every second.

b) Calculate the total kinetic energy of the air that arrives at the turbine every second.

c) Only 40% of the total kinetic energy of the wind can be converted into electrical energy.

Calculate the electrical power output of the wind turbine.

Solution

a) The air volume incident on the turbine in one second is equivalent to the volume of a cylinder of cross-sectional area $65\,\text{m}^2$ and length $15\,\text{m}$: $65\,\text{m}^2 \times 15\,\text{m} = 975\,\text{m}^3$.

To calculate the mass of air arriving in one second:

mass = density $\times$ volume = $1.3\,\text{kg m}^{-3} \times 975\,\text{m}^3 = 1270\,\text{kg}$.

b) The kinetic energy of this air is $\frac{1}{2} \times 1270 \times 15^2 = 140\,\text{kJ}$

c) 0.4 (40%) of this energy can be transferred to the electrical output = $57\,\text{kW}$.

> **Assessment tip**
>
> Analyse and describe energy transfers in a generating system carefully. Topics 2, 5 and 11 (HL) have direct relevance here. It may be worth reviewing the language of energy transfer in Topic 2.3.

> **Pumped storage hydroelectric systems** use electrical energy to pump water to a higher reservoir. The water stores gravitational potential energy. Then, at times of peak demand for electrical energy, the water runs back through the pump. The pump motor is now in reverse, so works as a dynamo, generating electricity. The energy transfer is from gravitational potential to kinetic to electrical.

> **Assessment tip**
>
> Scientists continue to research storage-battery technology. You should be aware of new types of energy storage and generation techniques as they develop.

SAMPLE STUDENT ANSWER

a) The hydroelectric system has four 250 MW generators. The specific energy available from the water is $2.7\,\text{kJkg}^{-1}$. Determine the maximum time for which the hydroelectric system can maintain full output when a mass of $1.5 \times 10^{10}\,\text{kg}$ of water passes through the turbines. [2]

This answer could have achieved 2/2 marks:

$1.5 \times 10^{10} \times 2.7 \times 10^3 = 4.05 \times 10^{13}\,\text{J}$

$4 \times 250 \times 10^6 = 1 \times 10^9\,\text{Watts}$

$\dfrac{4.05 \times 10^{13}}{1 \times 10^9} = 40500 \approx 41000\,\text{seconds}$

▲ A well-presented, correct answer, including a clearly-shown answer given to 2 significant figures.

b) Not all the stored energy can be retrieved because of energy losses in the system. Explain one such loss. [1]

This answer could have achieved 0/1 marks:

High temperature will cause energy loss in thermal energy.

▼ There appears to be the suggestion that high temperatures lead to energy loss. There are certainly frictional losses in the turbine bearings and resistive losses in the electrical cables. But this answer is too vague for credit.

8.2   THERMAL ENERGY TRANSFER

You must know:

✔ the three methods of thermal energy transfer: convection, conduction, radiation

✔ the solar constant is the amount of energy arriving at the Earth's orbit from the Sun

✔ the nature of black-body radiation

✔ what is meant by emissivity and albedo

✔ the albedo of the Earth's surface varies daily and seasonally

✔ what is meant by the greenhouse effect.

You should be able to:

✔ sketch and interpret a graph of the variation of intensity with wavelength for bodies emitting thermal radiation at various temperatures

✔ solve problems involving the use of Wien's displacement law and the Stefan–Boltzmann law

✔ describe the effects of the Earth's atmosphere on the mean surface temperature

✔ solve problems involving albedo, emissivity, solar constant and the Earth's average temperature.

Conduction occurs through collisions between electrons and atoms and through intermolecular interaction.

Convection occurs in fluids. The hot areas of a fluid are less dense than the cold areas, so the particles rise from the hot areas to the cold areas. The denser, cold areas then fall into the hot areas, creating a convection current.

Thermal radiation is emitted as electromagnetic waves by all objects at temperatures greater than 0 K (absolute zero). Thermal energy passes through a vacuum.

Thermal energy is transferred by three processes.

Conduction: The principal mechanism in solids.

Convection: The principal mechanism in fluids.

Radiation: The only mechanism for transfer in a vacuum.

All atoms possess kinetic energy at temperatures above absolute zero. An atom moving about its fixed position can transfer energy to other atoms close by when it has a greater energy (temperature) than its neighbours. In this way, energy can be transferred from high-temperature regions of a solid to low-temperature regions.

Example 8.2.1

Energy loss from a house can be reduced by placing foam between the inner and outer house walls (called cavity wall insulation).

Explain how heat transfer from inside a warm house to the cold exterior can be reduced by:

a) a cavity wall with no foam

b) a cavity wall with foam.

Solution

a) Air is a good insulator, so energy transferred through the inner wall is not conducted through the cavity.

b) A foam is a mixture of gas and solid. The gas is fixed in position within the solid and cannot convect. Both the solid and gas parts of the foam are poor conductors and do not allow effective energy transfer.

All objects radiate electromagnetic radiation. They also absorb electromagnetic radiation incident on them. A *black body* absorbs all the radiation incident on it. A black body also emits radiation with a pattern characterised by its temperature; this is called *black-body radiation*. The extent to which an emitter is imperfect compared with a black body is described by its *emissivity*.

> **Black-body radiation** is a continuous spectrum that depends only on the temperature of the radiator.
>
> **Emissivity** $e = \dfrac{\text{energy radiated from the surface of an object}}{\text{energy radiated from a black body at the same temperature and viewing conditions}}$
>
> For a black body, $e = 1$.
>
> For a perfectly reflecting and non-radiating object, $e = 0$.
>
> The fourth-power dependence of the total emitted power on temperature is shown in the graph above. Doubling the temperature from 3000 K to 6000 K means that the emitted power is increased by a factor of 16.

> **Wien's displacement law** predicts the characteristic peak in the graph of *relative intensity* against *wavelength*:
>
> $$\lambda_{max} \text{ (in m)} = \frac{2.90 \times 10^{-3}}{T \text{ (in K)}}$$
>
>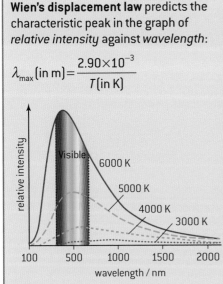
>
> The wavelength peak shifts to shorter wavelengths as the temperature of the black body increases.
>
> The area under the curve is given by the **Stefan-Boltzmann law**. This predicts that the total power P output by a black body radiator is
>
> $$P = e \, \sigma \, A T^4$$
>
> where
>
> σ is the Stefan–Boltzmann constant $(5.67 \times 10^{-8} \, \text{W m}^{-2} \, \text{K}^{-4})$
>
> e is the emissivity (included for bodies that are not perfect black bodies).

The Earth's climate is affected by many factors, including the incident and emitted radiation at the surface, and the composition of the atmosphere. Global warming and climate change modelling are important areas of research.

Radiation arrives at the Earth's surface from the Sun. The spectrum of this black-body radiation is determined by the temperature of the Sun (about 5700 K). A power of about 1400 W m^{-2} is delivered to the top of the atmosphere. The atmosphere is effectively transparent to most of the peak wavelength although there is some scattering of the shorter wavelengths.

Roughly 70% of this energy is absorbed by the Earth itself – the rest is reflected. The proportion reflected is the *albedo*.

The Earth then re-emits radiation because it is at a temperature greater than 0 K. For the average temperature of the planet to be constant, there must be a dynamic equilibrium between the incident and emitted radiation powers.

However, the Earth's temperature is much lower than that of the Sun and the peak wavelength emitted by Earth is in the infrared region—radiation with a longer wavelength. Certain atmospheric gases (called *greenhouse gases*) are opaque to infrared wavelengths and so energy is trapped inside the Earth–atmosphere system. This contributes to an increased average temperature for the Earth. This overall warming of the system through atmospheric absorption is known as the *greenhouse effect*.

When the concentrations of these gases increase in the atmosphere, increased amounts of outgoing infrared are absorbed, and more energy remains in the system. The overall albedo becomes smaller. This increases the temperature of the planet, which affects climate and sea level.

The effect of the increased concentration of greenhouse gases is known as the *enhanced greenhouse effect*.

Temperature of the Earth if it were a black body	278 K
Temperature of the Earth with an albedo of 0.3	255 K
Actual temperature of the Earth, including albedo and the greenhouse effect	287 K

> The **greenhouse gases** are water vapour (H_2O), carbon dioxide (CO_2), methane (CH_4) and dinitrogen monoxide (sometimes called nitrous oxide) (N_2O).

>> **Assessment tip**

The greenhouse effect is vital to ensure the average temperature of the Earth is high enough to sustain life. However, the **enhanced** greenhouse effect increases the average temperature of the Earth and causes significant climate change. Be clear about the distinction.

Climate modelling includes the factors discussed here and many more. The atmosphere–ocean system is extremely complex, and many national and international research groups are focusing their efforts on the implications of atmospheric change for the Earth's climate.

Albedo is $\dfrac{\text{total scattered power}}{\text{total incident power}}$.

The global annual mean albedo for Earth is 0.3. This means that 70% of the incident power from the Sun is absorbed by the planet.

The value for the albedo shows both daily and seasonal variations for any one location. The latitude of the location is also important as this determines the angle of the Sun in the sky. Cloud coverage also affects the value of albedo.

SAMPLE STUDENT ANSWER

The following data are available for a natural gas power station that has a high efficiency.

Rate of consumption of natural gas	$= 14.6\,\text{kgs}^{-1}$
Specific energy of natural gas	$= 55.5\,\text{MJkg}^{-1}$
Efficiency of electrical power generation	$= 59.0\,\%$
Mass of CO_2 generated per kg of natural gas	$= 2.75\,\text{kg}$
One year	$= 3.16 \times 10^{7}\,\text{s}$

a) Calculate, with a suitable unit, the electrical power output of the power station. [1]

This answer could have achieved 0/1 marks:

> $\dfrac{55.5}{14.6} = 3.8\,\text{MJkg/s at } 59\%$
>
> $= 2.2\,\text{MJkg/s}$

▼ It is important in this type of calculation to present your work carefully with a full explanation. The first step is to calculate the rate at which energy is converted. This is the *rate of gas consumption × specific energy of the gas* = 55.5 × 14.6. 59% of this energy is eventually transferred into an electrical form. This answer divides rather than multiplies by the specific energy. The presence of the unit (kg) is also confusing.

b) Calculate the mass of CO_2 generated in a year assuming the power station operates continuously. [1]

This answer could have achieved 1/1 marks:

> $(14.6 \times 2.75) \times (3.16 \times 10^{7}) = 1268740000\,\text{kg}$

▲ Every kilogram of gas burned in one second gives rise to 2.75 kg of CO_2. This answer evaluates this and converts (in one step) to the mass of CO_2 produced in one year. There are, however, too many significant figures in the answer; three significant figures would have been best.

c) Explain, using your answer to ii), why countries are being asked to decrease their dependence on fossil fuels. [2]

This answer could have achieved 0/2 marks:

> With the huge amount of carbon emissions that these power stations produce, the issue of disposal of the emissions arise without every affordable or environmentally efficient options.

▼ The answer should focus on an international perspective for the reduction of CO_2 release in the world. The consideration of 'disposal of the emissions' has no meaning. The answer should have considered the global need to reduce the emissions that arise from CO_2 release.

d) Describe, in terms of energy transfers, how thermal energy of the burning gas becomes electrical energy. [2]

This answer could have achieved 0/2 marks:

> Heat causes molecules to increase motion. Increased motion of electrons causes increase of electrical energy.

▼ The question requires a discussion of the transfers that take place in a gas-fired power station (the key phrase in the question is 'energy transfers'). This incorrect microscopic view of the transfer of electron energy is not appropriate.

Practice problems for Topic 8

Problem 1

A 250 MW generating station is to provide energy for a large and isolated town. Residents are to choose between a nuclear fission station and a coal-fired station. Both stations have lifetimes of about 25 years.

Compare the relative costs and the environmental impact of both types of generating station.

Problem 2

Water falls through a height of 4.8 m in a hydroelectric power station to provide electricity for a village.

a) Calculate the change in potential energy of a 1.0 kg mass of water falling through a vertical height of 4.8 m.

b) Discuss factors that affect the usefulness of hydroelectric power stations for electricity production.

Problem 3

When the concentration of carbon dioxide in the atmosphere doubles, the albedo of the Earth increases by 0.01.

Average intensity received at Earth from the Sun = 340 W m^{-2}

Average albedo = 0.30

a) Determine the change in the intensity of the radiation being reflected into space by the Earth.

b) State **one** reason why the answer to part a) is an estimate.

Problem 4

a) Coal-fired power stations emit greenhouse gases.

Outline what is meant by a *greenhouse gas*.

Problem 4 (continued)

b) The maximum power output of a coal-fired power station is 2.3 GW.

The energy density of nuclear fuel is 82 TJ kg^{-1}.

Determine the minimum mass of fuel that would be required by a nuclear power station to provide the same maximum annual energy output as the coal-fired station.

Problem 5

Solar cells are to provide the electrical energy for a small village with 29 houses. Each house uses an average power over one year of 800 W.

The intensity of solar radiation at the surface of the Earth is 650 W.

The efficiency of the conversion of solar energy to electrical energy is 15%.

a) Estimate the total area of solar cells needed to provide the power for the village.

b) State **one** reason why the area covered by solar cells will need to be greater than your estimate.

c) Suggest further problems that may occur when using only solar cells to provide the energy for the village.

Problem 6

The solar intensity arriving from the Sun at the radius of the Earth's orbit is 1400 W m^{-2}.

Mean radius of the Earth's orbit around the Sun = 1.5 × 10^{11} m

Radius of the Sun = 7.0 × 10^{8} m

a) Estimate the total output power of the Sun.

b) Use your estimate in part a) to deduce the temperature of the Sun.

9 WAVE PHENOMENA (AHL)

9.1 SIMPLE HARMONIC MOTION

You must know:

✔ the defining equation for simple harmonic motion (shm)

✔ how simple harmonic motion arises in the context of the simple pendulum and the mass–spring system

✔ how to solve problems about energy transfer in shm.

You should be able to:

✔ identify energy changes in shm

✔ describe how energy moves between kinetic and potential forms in shm

✔ solve problems involving displacement, velocity and acceleration during shm using graphs and algebra.

In Topic 4.1 you were introduced to the equation that defines simple harmonic motion:

acceleration of object = −(angular velocity)² × displacement of object

Angular velocity (the constant ω) is related to the time period T by $\omega = \dfrac{2\pi}{T}$ and to the frequency of the oscillation f by $\omega = 2\pi f$.

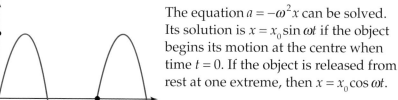

▲ **Figure 9.1.1.** Displacement–time graph for shm

The equation $a = -\omega^2 x$ can be solved. Its solution is $x = x_0 \sin \omega t$ if the object begins its motion at the centre when time $t = 0$. If the object is released from rest at one extreme, then $x = x_0 \cos \omega t$.

Figure 9.1.1 shows the graph of x against t. It is a sine (or cosine) wave.

For the cosine result, the graph begins at the maximum amplitude (x_0) and continues as a cosine graph.

The equation $v = \pm\omega\sqrt{\left(x_0^2 - x^2\right)}$ gives the variation with displacement rather than time. It has a $\pm$ sign because, at any single position between the ends of the motion, the object can be moving either towards or away from the centre.

>> **Assessment tip**

Notice that the displacement and acceleration equations satisfy the original shm definition because $x = x_0\sin(\omega t)$ and $a = -x_0\omega^2\sin(\omega t)$ so that $a = -\omega^2 x$, as required.

Example 9.1.1

A particle of mass m executes simple harmonic motion in a straight line with amplitude A and frequency f. Calculate the total energy of the particle.

Solution

This can be approached in a number of ways.

The maximum value of v occurs at the centre of the motion, $x = 0$.

So $v_{max} = A\omega$. As $\omega = 2\pi f$, $v_{max} = 2A\pi f$ and the kinetic energy $= \dfrac{1}{2}mv^2$ or $\dfrac{1}{2}m\,4\,A^2\pi^2 f^2$ or $2mA^2\pi^2 f^2$.

Knowing how displacement varies with time allows velocity–time and acceleration–time graphs to be drawn. These are connected through the gradients of the respective displacement and velocity graphs, as shown in the table.

		Object starts in centre of motion	Object starts at extremes of motion
Displacement	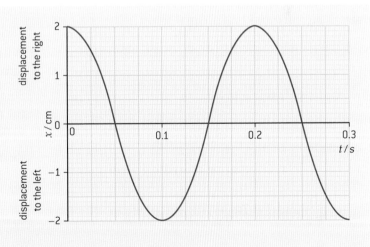	$x = x_0 \sin \omega t$	$x = x_0 \cos \omega t$
Velocity		$v = \omega x_0 \cos \omega t$ $v = \pm \omega \sqrt{\left(x_0^2 - x^2\right)}$	$v = -\omega x_0 \sin \omega t$ $v = \pm \omega \sqrt{\left(x_0^2 - x^2\right)}$
Acceleration		$a = -x_0 \omega^2 \sin \omega t$ $a = -\omega^2 x$	$a = -x_0 \omega^2 \cos \omega t$ $a = -\omega^2 x$

Example 9.1.2

Particle P moves with simple harmonic motion. This graph shows the variation of the displacement x of P in the medium with time t.

a) Calculate the magnitude of the maximum acceleration of P.

b) Calculate its speed at $t = 0.12\,\text{s}$.

c) State its direction of motion at $t = 0.12\,\text{s}$.

Solution

a) $T = 0.20\,\text{s}$.

$$a_{max} = \left(\left(\frac{2\pi}{T}\right)^2 x_0 = 31.4^2 \times 2.0 \times 10^{-2}\right) = 19.7 \approx 20\,\text{m s}^{-2}$$

b) The displacement at $t = 0.12\,\text{cm}$ is $-1.62\,\text{cm}$.

$$v\left(= \frac{2\pi}{T}\sqrt{x_0^2 - x^2}\right) = 31.4\sqrt{\left(2.0 \times 10^{-2}\right)^2 - \left(1.62 \times 10^{-2}\right)^2} = 0.37\,\text{m s}^{-1}$$

c) To the right

In Topic 4.1 it was noted that the shapes of the energy–time graphs are not sine or cosine curves and that one cycle of energy variation has half the time of one cycle of shm.

The graph of E_k and E_p against displacement is parabolic:

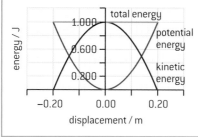

The equation for kinetic energy E_k is $\frac{1}{2}mv^2$. Combining this with both the displacement and time variants gives

$$E_k = \frac{1}{2}m\omega^2\left(x_0^2 - x^2\right) = \frac{1}{2}m\omega^2 x_0^2 \cos^2\left(\omega t\right).$$

For the potential energy in the system, the key is to recognise that, for true shm, no energy loss occurs and the total energy is constant at $E_T = \frac{1}{2}m\omega^2 x_0^2$. This must be the sum of E_k and E_p, so E_p is always

$$\frac{1}{2}m\omega^2 x_0^2 - E_k = \frac{1}{2}m\omega^2 x^2 = \frac{1}{2}m\omega^2 x_0^2 \sin^2\left(\omega t\right).$$

Two contexts for simple harmonic motion are the simple pendulum and the mass–spring system.

Simple pendulum

The simple pendulum is an example of approximate shm. This is because there is a simplification in the mathematics, meaning that the motion is only simply harmonic for small swings.

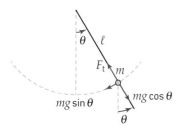

The force acting on the pendulum bob to return it to the centre is $mg \sin\theta$ and this is equal to ma.

So, $g\sin\theta = a$ and, for small angles $< 10°$ $\sin\theta \approx \theta$ (in rad).

However, $\theta = \dfrac{x}{l}$ leading to $g\dfrac{x}{l} = -a$ (the negative sign because x is measured away from the vertical and a is towards the vertical).

Thus $a = -\left(\dfrac{g}{l}\right)x$. Comparing this with $a = -\omega^2 x$ shows that $\omega = \sqrt{\dfrac{g}{l}}$ and $T = 2\pi\sqrt{\dfrac{l}{g}}$.

Mass–spring system

Providing that the spring is elastic and its extension is always directly proportional to the force acting on it, the motion of the mass–spring system is exactly simple harmonic.

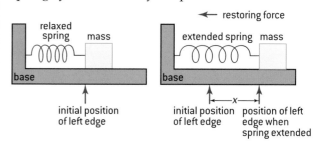

The force F acting on a spring is directly proportional to its extension x: $F = kx$.

When the spring is extended, a force acts to return the mass to the equilibrium position.

So, $ma = -kx$ (the negative sign because the positive direction is to the right in the diagram but the force acts to the left).

Rearranging gives $a = -\left(\dfrac{k}{m}\right)x$ which matches the shm equation, giving

$$\omega = 2\pi\sqrt{\dfrac{m}{k}}.$$

Example 9.1.3

The graph shows the variation of total potential energy with time for a mass–spring system. The mass of the spring is 0.32 kg and the maximum kinetic energy in the system is 20 mJ.

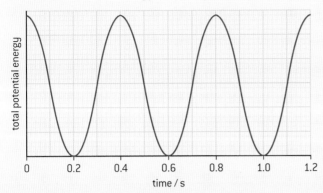

a) State, with a reason, the time period of oscillation of the mass on the spring.

b) Calculate the spring constant k of the spring used.

c) Determine the amplitude of the oscillation.

Solution

a) The potential energy cycles twice for one time period of the system. The time period of the mass on the spring is 0.80 s.

b) $T = 2\pi\sqrt{\dfrac{0.32}{k}}$ and so $k = \dfrac{4\pi^2 \times 0.32}{(0.80)^2}$; $k = 20\,\mathrm{N\,m^{-1}}$

c) $2.0 \times 10^{-2} = \dfrac{1}{2}m\omega^2 x_0^2 = \dfrac{1}{2}0.32 \times \left(\dfrac{2\pi}{0.80}\right)^2$ which gives $x_0 = 0.045\,\mathrm{m}$.

SAMPLE STUDENT ANSWER

A small ball of mass m is moving in a horizontal circle on the inside surface of a frictionless hemispherical bowl.

a) The ball is now placed through a small distance x from the bottom of the bowl and is released from rest.

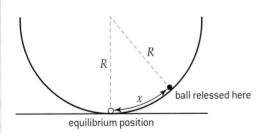

equilibrium position

The magnitude of the force on the ball towards the equilibrium position is given by

$\dfrac{mgx}{R}$

where R is the radius of the bowl.

Outline why the ball will perform simple harmonic oscillations about the equilibrium position. [1]

This answer could have achieved 1/1 marks:

The acceleration of the ball is proportional to displacement and is directed towards the equilibrium position.

▲ This answer captures both essential points about the definition of shm: the magnitude and the direction.

b) The radius of the bowl is 8.0 m.
Show that the period of oscillation of the ball is about 6 s.   [2]

This answer could have achieved 2/2 marks:

> ▲ This is a good answer. The acceleration of the sphere is deduced (and is shown as ∝x) thus identifying the constant of proportionality. This is ω^2 and leads to a correct calculation.

$$F = \frac{mgx}{R} \qquad a = \frac{x}{R} \qquad \frac{g}{R} = \omega^2$$

$$\omega = -\sqrt{\frac{g}{R}} = 1.1 s^{-1}$$

$$T = \frac{2\pi}{\omega} = \frac{2\pi}{1.1 s^{-1}} = 5.75 \approx 6s$$

9.2   SINGLE-SLIT DIFFRACTION

You must know:

✔ the nature of single-slit diffraction

✔ how to determine the position of the first minimum in a diffraction pattern.

You should be able to:

✔ describe the appearance of a single-slit diffraction pattern produced by monochromatic light and by white light

✔ describe the effect of changing slit width on the appearance of a diffraction pattern.

> **» Assessment tip**
>
> θ must be in radians for the equation $\theta = \dfrac{\lambda}{b}$. Ensure that your calculator is set correctly. This advice applies to all angle calculations in DP physics.

> Topic 9.3 provides advice about precision in drawing the intensity–position graph for diffraction.

> **» Assessment tip**
>
> All wave types demonstrate diffraction; you could be asked about diffraction in sound waves or microwaves, and so on. The basic physics is unaltered.

The **first minimum position** of a single-slit diffraction pattern is given by $\theta = \dfrac{\lambda}{b}$, where θ is the angle between the central maximum and the first minimum, λ is the wavelength and b is the distance from the single slit to the observing screen.

The proof of this equation uses the approximation $\sin\theta \approx \theta$ and so the equation is only true for small angles.

Example 9.2.1

Sound waves of wavelength 35 cm are incident on a gap in a fence of width 2.7 m.

a) The first minimum in the intensity of the sound is at an angle of θ from the central maximum.

     Calculate θ.

b) The frequency of the sound is reduced without changing its amplitude.

     State and explain how this will affect the position of the first minimum.

Solution

a) $\theta = \dfrac{\lambda}{b} = \dfrac{0.35}{2.7}$ rad, leading to 7.4°.

b) A reduced frequency means a greater wavelength, so the sound will be diffracted through a larger angle.

Monochromatic light, diffracted by a slit, gives the instantly recognizable diffraction pattern. At HL, you should be able to sketch the diffraction patterns with precision and confidence. You should also be able to carry out straightforward calculations of the position of the first minimum position.

🔗 Some situations you meet in the course have circular apertures rather than single slits. There is an additional factor in the diffraction equation which becomes $\theta = \dfrac{1.22\lambda}{b}$ when the aperture is circular.

SAMPLE STUDENT ANSWER

Microwaves of wavelength 32 mm leave a transmitter through an aperture of width 64 cm.

Estimate the angle, in degrees, between the central maximum and the first minimum of the diffraction pattern. [2]

This answer could have achieved 1/2 marks:

$$\theta = \frac{\lambda}{b} = \frac{32}{640} = 0.05°$$

▼ The answer has not been converted from radians into degrees. The answer should be 2.8°.

>> **Assessment tip**

Some problems ask for the pattern differences between red and blue in the context of a white-light source. The best way to answer these is from first principles:

$$\theta = \frac{\lambda}{b}$$

Another way to remember the effect is that in refraction, red light is refracted *less* than blue light. In diffraction red is diffracted *more* than blue.

9.3 INTERFERENCE

You must know:

✔ what is meant by Young's double-slit experiment

✔ how to investigate the Young's double-slit (two-slit interference) arrangement experimentally

✔ the appearance of the two-slit interference pattern and how it is modified (modulated) by diffraction of the waves at each of the slits

✔ the interference patterns from to multiple slits

✔ the interference pattern from a diffraction grating

✔ what is meant by thin-film interference and how to solve problems including thin-film interference.

You should be able to:

✔ describe two-slit interference patterns, including the modulation caused by single-slit diffraction

✔ sketch and interpret intensity–position graphs of two-slit interference patterns

✔ solve problems involving the diffraction-grating equation

✔ describe the conditions required for constructive and destructive interference occurring in thin films and at their interfaces, including the phase changes that occur at the interface and the effect of refractive index.

Here's how to investigate the optical behaviour of a pair of slits.

• Take your pair of slits and measure the distance d between them. You could use a microscope simultaneously imaging the slits and a metal ruler to do this.

• Project the light from a laser pointer (taking care not to look at the direct beam or a reflection of it) through your double slit and onto a screen at least 3 m away. Measure the distance from the slits to the screen—this is D.

🔗 The double-slit interference pattern was introduced in Topic 4.4. The basic equation developed there was $s = \dfrac{\lambda D}{d}$. In Topic 9.3, we look at how the slits themselves modify the basic double-slit pattern.

Topic 9.3 provides advice about precision in drawing the intensity–position graph for diffraction.

The Young's slit *interference* pattern (each fringe of equal intensity) is **modulated** by a single-slit *diffraction* pattern (see Figure 9.3.1). The diffraction pattern acts as an envelope for the interference fringe pattern, setting a maximum limit on the intensity at any position.

- You should see a fringe pattern on the screen. Use a ruler to measure the distance between a known number of fringes (10–20 of them is about right). Use this measurement to determine s, the distance between adjacent fringes.

- Use the data to calculate the wavelength of the laser.

So far interference has been treated as occurring at slits that are infinitely thin. This is a poor model for interference by multiple slits as both interference and diffraction must occur at the slits.

Assume that a double-slit experiment has two slits of the same (finite) width separated by a small distance. Both slits give rise to identical diffraction patterns that are offset by the slit separation. It is the interference of these two diffracted beams that gives rise to the final fringe pattern.

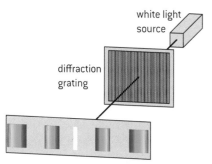

▲ Figure 9.3.2. The diffraction grating arrangement

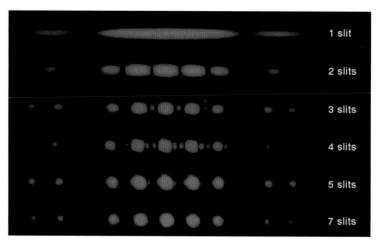

▲ Figure 9.3.1. A single-slit diffraction pattern modulates the interference fringes

As the number of slits increases to three or more, with the individual slit width and spacing unchanged, then other effects appear. The fringes become sharper. Subsidiary maxima appear between the fringes but, as the number of slits increases, these become relatively less intense than the main peaks between them.

When the number of slits becomes very high the arrangement becomes that of a *diffraction grating*. The pattern for monochromatic light becomes a bright central maximum surrounded by intense sharp lines with darkness between them. These lines are *orders* and are given integer labels counting out from the centre which is zero. With many wavelengths, the individual lines become spectra—lines or a continuous red to violet band (Figure 9.3.2).

The **order** of a spectrum is n. The wavelength of a line in the spectrum is λ. The distance between adjacent slits, the slit separation, is d.

For a diffraction grating, the angle θ between the straight-on beam and a line is given by $n\lambda = d\sin\theta$.

The **number of slits per metre** is how diffraction gratings are often specified $= \dfrac{1}{d}$.

Example 9.3.1

A diffraction grating has 4.5×10^5 lines m^{-1}. Light of wavelength 486 nm is normally incident on the grating. Determine the highest order diffracted image that can be produced for this wavelength by the diffraction grating.

Solution

The highest order occurs at $\theta = 90°$. Using this value in the equation

gives $n = \dfrac{d\sin\theta}{\lambda} = \dfrac{2.22 \times 10^{-6} \times \sin(90°)}{4.86 \times 10^{-7}} = 4.6$

This means that the fourth order is the highest order that can be observed.

When a wave pulse travels to the end of a string, conditions at the boundary determine the shape of the reflected pulse. Such effects in strings are easy to imagine. Less easy are phase changes that occur when light travels between two media.

The reflected and refracted waves of light interfere with each other and the associated phase changes play their part in determining what is reflected and transmitted.

The phase changes at boundaries depend on the refractive indices of the two media.

When the electromagnetic wave is reflected at an optically denser medium (higher refractive index), there is a phase change of π rad ($\equiv 180°$).

When the wave is reflected at an optically less dense medium, there is no phase change.

The phase changes are summarised in Figure 9.3.3.

There are two cases that illustrate the effect.

① A camera lens can be made anti-reflective *at one wavelength* by coating the front surface with a transparent substance that has a refractive index midway between that of the air and that of the glass.

There is a phase change of π at both X and Y; the refractive index increases at both interfaces. When the thickness of the coating is $\frac{\lambda}{4}$, the light from Y has travelled an extra distance $\frac{\lambda}{2}$ (there and back). In phase change terms, there is π ($\equiv 180°$) difference between the ray at X and the ray at Z.

When the real thickness of the coating is d, the effective thickness is $d \times n$, where n is the refractive index. The light, therefore, travels a total XZ distance of $2nd$, and when this is equal to $\frac{\lambda}{2}$ (or odd multiples of it) the light from Z and the light from X are π ($\equiv 180°$) out of phase. The condition for destructive interference is $2nd = \left(m + \frac{1}{2} \right)\lambda$, where m is an integer.

All the reflection from the front surface of the lens is eliminated at one specific wavelength.

② Patterns of coloured fringes can sometimes by seen in the oil layer on a wet road (Figure 9.3.4).

The relationship between the refractive indices of the layers is different from the lens coating case. The oil has a larger refractive index than the water beneath it. Now there is a phase shift at the top surface (air–oil) but not at the bottom reflection (oil–water). When the film has a thickness of $\frac{\lambda}{2}$ then the total path difference in the oil is λ and this will lead to destructive interference between the ray from X and the ray from Z. This time, the destructive interference case is $2nd = m\lambda$ for light at normal incidence.

The light appears coloured because one wavelength is missing. As you look at the film different colours are removed at different angles so you see a multi-coloured fringe pattern.

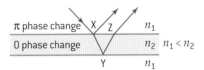

Boundary conditions were covered in Topic 4.5 for fixed and free boundaries, and for those where two strings of different mass per unit lengths are joined.

▲ **Figure 9.3.3.** Phase changes at interfaces

>> **Assessment tip**

In these diagrams the incident, reflected and refracted rays are drawn well away from the normal to make things clear. In reality, the rays lie almost on top of one another.

>> **Assessment tip**

You may find it odd that cancellation of the reflected light means that more light enters the lens. Remember that light is an electromagnetic wave and our model is a simplistic one. As usual, energy must be conserved. When none is reflected from the surface then, allowing for energy loss in the coating and glass, overall more must be able to pass through the lens system.

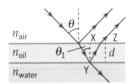

▲ **Figure 9.3.4.** Interference at an oil film on water

>> **Assessment tip**

The data booklet only gives one set of equations for thin-film interference and there are two possible sets depending on the layer refractive indices. It is best to learn how to solve individual cases rather than rely on the equations—just use these as a memory aid.

SAMPLE STUDENT ANSWER

A student investigates how light can be used to measure the speed of a toy train.

Light from a laser is incident on a double slit. The light from the slits is detected by a light sensor attached to the train.

The graph shows the variation with time of the output voltage from the light sensor as the train moves parallel to the slits. The output voltage is proportional to the intensity of light incident on the sensor.

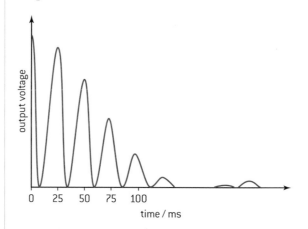

direction of travel

double slit

laser light → | 1.5 mm

toy train

light sensor

5.0 m

not to scale

a) Explain, with reference to the light passing through the slits, why a series of voltage peaks occurs. [3]

This answer could have achieved 3/3 marks:

▲ An answer that shows a good understanding of the unusual context. The train is moving through an interference pattern and the question asks for an *explanation* of the fringes in terms of path differences, and a *link* between the voltage minima and the intensity minima. This answer gives an account of both of these.

When light wave pass through the slit, it creates two wave source with same frequency. In some position along the train track, the path length difference between the two source is integer value of the wavelength λ, which creates constructive interference, therefore peaks are created. At some other places path length difference is $n\dfrac{\lambda}{2}$, so the wave interfere destructively, creating a place with no intensity, therefore no voltage.

b) i) The slits are separated by 1.5 mm and the laser light has a wavelength of 6.3×10^{-7} m. The slits are 5.0 m from the train track. Calculate the separation between two adjacent positions of the train when the output voltage is at a maximum. [1]

This answer could have achieved 1/1 marks:

▲ The question continues as a routine calculation involving the double slit equation. The answer is correct.

$$S = \frac{(6.3 \times 10^{-7})(5)}{1.5 \times 10^{-3}} = 0.0021\,m$$

(ii) Estimate the speed of the train. [2]

This answer could have achieved 2/2 marks:

▲ This problem involves linking the rate at which the train passes the fringes (from the graph) to the distance between fringes in (b)(i). The solution is correct.

$$\frac{0.0021}{25 \times 10^{-3}} = 0.084\,ms^{-1}$$

9.4   RESOLUTION

You must know:

✔ the meaning of diffraction-aperture size

✔ what is meant by resolution in the context of single-slit diffraction and diffraction gratings

✔ the definition of the resolvance of a diffraction grating.

You should be able to:

✔ solve problems involving resolution and resolvance of diffraction gratings

✔ state and explain the Rayleigh criterion for the resolution of an image of two slits diffracted by a single slit or circular aperture.

You see a light on a hill at night a long distance away. Can you decide whether this is one lamp or two lamps close together? This question was asked by the British scientist, Lord Rayleigh. Rayleigh's idea was that two images can just be resolved when the central maximum of one diffraction pattern coincides with the first minimum of the other. As Figure 9.4.1(b) shows, at this separation, there is a small dip in the total intensity. But in Figure 9.4.1(a) the patterns are closer than the Rayleigh criterion and the summed intensity pattern cannot be distinguished from that of a single pattern. In Figure 9.4.1(c) the patterns are well apart and there is a pronounced dip.

(a)

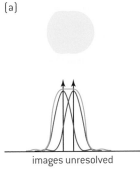

images unresolved

Example 9.4.1

Two vertical straight filaments of lamps are separated by 5.0 mm. An observer views the filaments through a narrow slit from a distance of 2.5 m. The filaments are just resolved with yellow light of wavelength 580 nm.

a) Determine, in degrees, the minimum angle of resolution using the slit and the blue light.

b) Deduce the slit width.

c) State and explain whether the filaments will be resolved when viewed at the same position using the same slit and a red filter.

Solution

a) Geometrically, the minimum resolved angle is

$$\theta = \frac{\text{distance between filaments}}{\text{distance to filaments}} = \frac{5 \times 10^{-3}}{2.5} = 2 \times 10^{-3} \text{ rad.}$$

This answer needs to be converted to degrees; 2×10^{-3} rad $\equiv 0.11°$

b) Using the Rayleigh criterion and rearranging,

$$b = \frac{\lambda}{\theta} = \frac{580 \times 10^{-9}}{2.0 \times 10^{-3}} = 0.29 \times 10^{-3} \text{ m.}$$

The slit width is 0.29 mm.

c) Red light has a longer wavelength than yellow light. This will mean a wider diffraction angle and so the diffraction patterns will overlap more. The 'dip' in the total intensity will disappear and the images will not be resolved.

(b)

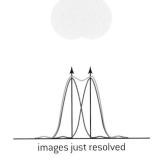

images just resolved

(c)

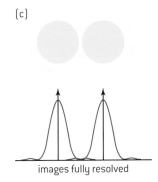

images fully resolved

▲ Figure 9.4.1. Resolution (a) not resolved (b) just resolved (c) well resolved

➤➤ Assessment tip

When writing about resolution, remember that it is the *image* that is resolved, not the *object*. Resolution is a property of an optical or other image-forming system.

There is a question of resolution when using a diffraction grating; for example, when viewing spectral lines within the same order. Using a coarse grating with slits well separated means that two similar wavelengths will merge. With a fine grating, the two wavelengths will be spread out in the grating spectrum enough to be distinguished. The ability of a grating to spread spectral lines in this way is called its *resolvance*.

> **Resolvance** $R = \dfrac{\lambda}{\Delta\lambda} = mN$, where
>
> m is the order of the spectrum and N is the total number of slits illuminated.
>
> For the examination you can use either of the λ values or the mean λ.
>
> The **resolvance** of a diffraction grating is the reciprocal of the fractional difference in wavelengths between two wavelengths that can just be resolved; in other words .
>
> $\dfrac{1}{\left(\dfrac{\Delta\lambda}{\lambda}\right)} = \dfrac{\lambda}{\Delta\lambda}$

Example 9.4.2

Two violet lines in the hydrogen spectrum have wavelengths 434 nm and 410 nm. These lines are just resolved in the second-order spectrum by a diffraction grating with 30 lines mm^{-1}.

Estimate the width of the light that is incident on the grating.

Solution

$\Delta\lambda = 24$ nm and $\lambda = 410$ nm so $N = \dfrac{\lambda}{m\Delta\lambda} = \dfrac{410}{2 \times 24} = 8.5$

It is best to round up to 9 at this point; this means that the light incident on the grating must be illuminating nine slits and is, therefore, $\dfrac{9}{30} = \dfrac{3}{10} = 0.3$ mm wide.

Notice that using 434 nm changes the number of slits to 9(.05). So, in terms of an estimate, this is not an important difference.

SAMPLE STUDENT ANSWER

A diffraction grating is used to resolve two lines in the spectrum of sodium in the second order. The two lines have wavelengths 588.995 nm and 589.592 nm.

Determine the minimum number of slits in the grating that will enable the two lines to be resolved. [2]

This answer could have achieved 1/2 marks:

$\Delta\lambda = 1.597$ nm; mean $\lambda = 589.29$ nm

$N = 589.29 \div 1.597 \div 2 = 184$

> ▼ There is an error in the calculation of $\Delta\lambda$; this should be 0.597 nm. However, there is enough detail for the examiner to follow this error through. The rest of the answer is correct and scores 1.

9.5 THE DOPPLER EFFECT

You must know:

✔ that shifts in frequency are observed when there is relative motion between a source and an observer

✔ that there are differences between the Doppler effect as observed in light waves and in sound waves.

You should be able to:

✔ describe situations where the Doppler effect is useful and solve problems using the Doppler effect

✔ sketch diagrams showing how the Doppler effect occurs

✔ interpret wave diagrams given to you to explain the Doppler effect.

When a source of a sound wave or electromagnetic radiation is at rest relative to an observer, both source and observer agree about the frequency emitted by the source. A Doppler shift occurs in the frequency when the source and observer move relative to each other. The cases for sound and electromagnetic radiation are treated differently.

When the source is moving relative to the transmission medium (taken to be air here) and the observer is stationary, the effect is easy to imagine. The wavefronts from the source are emitted at a constant rate (the frequency) but the source 'catches up' with its own waves. These waves, once emitted, continue to move in the medium at the wave speed. They cross the position of the observer more quickly than they were emitted and so a higher frequency is detected.

For the source moving away, the distances between waves are 'stretched' compared with the stationary case and the observer detects a lower frequency than that emitted.

When the observer moves relative to a stationary source, the emitted wavefronts are not distorted by source movement and are concentric spheres—centred on the source. However, the observer crosses the wavefronts more quickly (when approaching) or more slowly (when receding) and so a Doppler effect is detected again.

This theory assumes that the source and observer are moving along the line between their centres. When this is not true, the speed components along this line must be used.

The data booklet gives two separate equations for the Doppler effect: one for a moving source $f' = f\left(\dfrac{v}{v \pm u_s}\right)$ and another for a moving

observer $f' = f\left(\dfrac{v \pm u_o}{v}\right)$. Here f' is the observed frequency and f is the

emitted (source) frequency. The speed of the sound is v and the speed of the observer is u_o with the source speed u_s. However, you need to know which sign to use.

Example 9.5.1

A whistle emits sound of frequency of 1000 Hz and is attached to a string. The whistle is rotated in a horizontal circle at a speed of 30 m s^{-1}. The speed of sound in air is 330 m s^{-1}.

An observer is standing a long way from the whistle on the same horizontal plane.

a) Explain why the sound heard by the observer changes regularly.

b) Determine the maximum frequency of the sound heard by the observer.

Solution

a) The whistle (source) moves first towards the observer and then away. When the whistle is moving towards the observer, the wavefronts are compressed and the rate at which they cross the observer increases compared with a stationary whistle. The frequency is increased. When source and observer are moving apart, the wavefronts are further apart and the frequency decreases. As the relative speed is constantly changing, the frequency heard changes regularly too.

b) The maximum frequency occurs when the source and observer are approaching.

$$f = f' = \frac{1000 \times 330}{(330 - 30)} = 1100\,\text{Hz}$$

> ## ≫ Assessment tip
>
> Work from first principles each time. Imagine you are standing by the roadside and a police car goes along the road at speed sounding its siren. The sound goes from a high to a low frequency, changing as the car draws level with you. In the moving source equation, the negative sign must be used for the approaching police car and the positive used for the receding police car.
>
> An alternative way to remember this is to memorise the full expression but only for the approaching case, this is
>
> $$f' = f\left(\frac{v + u_0}{v - u_s}\right).$$
>
> When either u_s or u_0 is zero, the term simply disappears from the equation.
>
> When the objects are moving apart, reverse the signs.

 Redshifts occur in Options A and D.

Option A

An object moving relative to another experiences **time dilation** which affects the perceived rate at which it emits pulses; this is the relativistic Doppler shift.

Radiation originating from a region of a large star or black hole has a **gravitational redshift** as energy is required to move it to infinity. This energy is gained at the expense of frequency ($E = hf$).

Option D

Light takes a long time to reach us from distant galaxies. The Universe has expanded during this time and so the wavelengths have stretched as a consequence. This is the **cosmological** or **galactic redshift**.

The Doppler effect with electromagnetic waves needs care and so DP Physics only treats cases where the observer and source speeds are much less than the speed of light.

With this assumption, $\dfrac{\Delta f}{f} = \dfrac{\Delta \lambda}{\lambda} \approx \dfrac{v}{c}$.

(At higher speeds, the equation is more complicated.)

When the source and observer are moving away from each other, the observed wavelength is longer than that emitted, so the frequency is reduced. This is a redshift. When the source and observer are moving closer together, there is a blueshift.

The Doppler effect has many applications. It can be used with radar to provide the speeds of moving objects, and in weather forecasting. In radar astronomy, microwaves are reflected from objects such as the Moon. For a reflection, the frequency shift Δf is $\approx 2f \dfrac{v}{c}$.

SAMPLE STUDENT ANSWER

Police use radar to detect speeding cars. A police officer stands at the side of the road and points a radar device at an approaching car. The device emits microwaves which reflect off the car and return to the device. A change in frequency between the emitted and received microwaves is measured at the radar device.

The frequency change Δf is given by

$$\Delta f = \frac{2fv}{c}$$

Where f is the transmitter frequency, v is the speed of the car and c is the wave speed.

The following data are available.

Transmitter frequency f = 40 GHz

Δf = 9.5 kHz

Maximum speed allowed = 28 ms⁻¹

a) Explain the reason for the frequency change. [3]

This answer could have achieved 0/3 marks:

As the car moves towards the police officer, the device's waves have an increased perceived frequency due to the relative motion of the car to the police officer.

$$f' = f\left(\frac{v}{v \pm u_s}\right)$$

As the reflected microwaves are reflected on the car, they travel back as if they had the additional speed of the car. As a result, the waves reach the device faster than they would have if they were reflected off a non-moving source, and therefore there is a frequency change.

▼ The answer begins well with a discussion of relative motion. However, towards the end there is a statement that the "waves reach the device faster than they would have". The waves are travelling at the speed of light and this does not change depending on the observer's speed. This major error in physics disqualifies the whole answer.

This answer could have achieved 2/3 marks:

This case is a Doppler effect where there is a moving source of waves and a stationary observer (receiver).
Because the source is moving, the distance that the waves travel is reduced, which causes the same waves to be 'squeezed' into a smaller distance. The wavelength is then decreased while the wavespeed remains the same. Therefore, frequency increases $(c = f\lambda)$.

▲ There is clear identification of the reason for the frequency shift. There is also the recognition that the frequency increases (this is not given in the question and is always worth stating if not).

▼ However, the idea that the distance the waves travel is reduced is incorrect – better to say that the wavefronts are closer because the source has moved since the emission of the previous wavefront.

b) Suggest why there is a factor of 2 in the frequency-change equation [1]

This answer could have achieved 0/1 marks:

Because the distance travelled by the wave is two times the distance between the source and the radar (car).

▼ The factor of 2 is because of the reflection from the moving object which acts first as observer and then as source. The echo distance idea here is wrong.

Practice problems for Topic 9

Problem 1
A crane moves a load of mass 1.5×10^3 kg that is supported on a vertical cable. The mass of the cable is negligible compared with the mass of the load. When the crane stops, the load and cable behave like a simple pendulum. The oscillation of the load has an amplitude of 5.0 m and period of 8.0 s.

a) Calculate the length of the cable that supports the load.

b) (i) Determine the maximum acceleration of the load as it swings.

(ii) Calculate the force on the load that produces the acceleration in b) ii).

Problem 2
Laser light with wavelength 6.2×10^{-7} m is incident on a single slit of width 0.15 mm.

a) Calculate, in degrees, the angle between the central maximum and the first minimum in the diffraction pattern.

b) Describe and explain the change in the appearance of the pattern when the monochromatic laser light is replaced by white light.

Problem 3
A diffraction grating with 10 000 lines m^{-1} is used to analyse light emitted by a source.

The source emits a range of wavelengths from 500 nm to 700 nm.

a) Calculate the angle from the central maximum at which the first-order maximum for the 500 nm wavelength is formed.

b) Calculate the angular width of the first-order spectrum.

c) A detector is positioned a distance of 2.0 m from the grating to detect the maxima. Calculate the distance between the extreme ends of the first-order spectrum in this position.

Problem 4
When white light is reflected from a thin oil film floating on water, a series of coloured fringes is seen.

a) Outline how the fringes arise.

b) State and explain the changes to the appearance of the oil film when it is viewed from different angles of incidence.

Problem 5
a) Outline what is meant by the Rayleigh criterion for resolution.

b) A radio telescope with a dish diameter of 120 m receives radio signals of wavelength 6.0×10^{-2} m.

Calculate the minimum angular separation between radio images which this telescope can just resolve.

Problem 6
A train horn emits a frequency f. An observer moves towards the stationary train at constant speed and measures the frequency of the sound to be f'. The speed of sound in air is 330 m s^{-1}.

a) Explain, using a diagram, the difference between f' and f.

b) The frequency f is 300 Hz. The speed of the observer is 15.0 m s^{-1}.

Calculate f'.

10 FIELDS (AHL)

10.1 DESCRIBING FIELDS

You must know:

- ✓ what is meant by a gravitational field and an electrostatic field

- ✓ what is meant by gravitational potential and electric potential

- ✓ that a field line indicates the direction of a field and that the density of field lines indicates the strength of a field

- ✓ what is meant by an equipotential surface

- ✓ that an object with mass can move on a gravitational equipotential surface without doing work

- ✓ that a charged object can move on an electric equipotential surface without doing work.

You should be able to:

- ✓ represent masses and charges, and electric and gravitational lines of force (field lines) using suitable symbols

- ✓ map fields using potential and describe the relationship between a field line and an equipotential surface

- ✓ describe the electric field around a single point charge, between two point charges, and between two charged plates

- ✓ describe the gravitational field around a point mass, around a spherical mass, and close to the surface of a large massive object.

≫ Assessment tip

Electrostatic fields and gravitational fields involve very similar concepts and vocabulary. Use your knowledge of one of these fields to reinforce your understanding of the other.

A field is a region in which a force acts 'at a distance' on a mass (gravitation) or a charge (electrostatics).

Field strength is defined as the $\dfrac{\text{force acting on a test object}}{\text{size of test object}}$.

	Electric field strength	Gravitational field strength
Definition	$E = \dfrac{F}{q}$	$g = \dfrac{F}{m}$
Nature of test object	small positive charge	small mass
Unit	$N\,C^{-1} \equiv V\,m^{-1}$	$N\,kg^{-1} \equiv m\,s^{-2}$

In both cases, the test object is small and does not change the field strength where it is placed.

Because a force acts on an object, work must be done to move the object in the field:

work done = force × distance

Potential is a quantity independent of the magnitude of mass or charge of the object on which the force acts. Potential is a property of the field, not of individual objects moving in it.

<table>
<tr><td>

Electric potential at a point is the work done W in moving unit positive charge from infinity to the point.

Electric potential is given by

$W = q \Delta V_e$

where q is the charge and ΔV_e is the electric potential difference.

Unit of electric potential: $J\,C^{-1} \equiv V$

</td><td>

Gravitational potential at a point is the work done W in moving unit mass from infinity to the point.

Gravitational potential is given by

$W = m \Delta V_g$

where m is the mass and ΔV_g is the gravitational potential difference.

Unit of gravitational potential: $J\,kg^{-1}$

</td></tr>
</table>

<table>
<tr><td>

The concept of **infinity** produces a 'standard' place where potential is zero. In both electric and gravitational fields, force varies as $\dfrac{1}{distance^2}$.

</td><td>

At very large distances (infinity) the force must be zero. The potential at infinity is zero.

</td></tr>
</table>

This topic picks up from Topics 5.1 and 6.2 which covered Coulomb's law and Newton's law of gravitation. You should understand the principles of these laws.

Equations introduced in the earlier topics include:

Coulomb's law $F = k\dfrac{q_1 q_2}{r^2}$;

Newton's law of gravitation

$F = G\dfrac{m_1 m_2}{r^2}$.

> ## Assessment tip

To help your understanding, link these new ways that describe energy change in a field to the meaning of potential difference in current electricity. This is explored in more detail in Topic 10.2.

An important distinction between gravitational and electric fields is that gravitational potential is always negative whereas electric potential can be either positive or negative.

To see this, first think about gravity. Gravitational force is always attractive. To move a test object to infinity from somewhere close to a mass means that work must be done (this is because an attraction has to be overcome). Energy is transferred to the system to move the mass. At infinity, the potential of the system is zero. Therefore, it must have been negative before the transfer to infinity took place.

In electric fields, when a positive test charge is attracted by a negative charge, the situation is the same as for gravity; the potential is negative. However, a positive charge and a positive test charge repel, and work has to be done to keep them together. When you want to regain the stored positive potential energy, simply release the two charges and they will fly apart.

Point charges, point masses, charged spheres and spheres with mass have radial fields associated with them (Figure 10.1.1).

Figure 10.1.2 shows the uniform field and the equipotentials between two parallel charged plates.

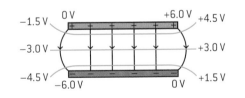

▲ Figure 10.1.2. Electric field line pattern for parallel plates

Between the plates the field lines are parallel and equally spaced; the field is uniform. Outside the plates, the field strength must fall away to the magnitude outside. The curved edge effects are the way in which the system makes this transition.

Although the Earth has a radial gravitational field, we live close to the surface so the separation of the lines is not apparent to us—broadly speaking, we live in a uniform gravitational field.

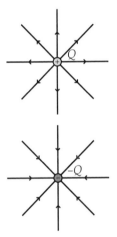

▲ Figure 10.1.1. Radial fields for positive and negative point charges

> ## Assessment tip

You can ignore the effect of edge effects in the DP Physics course but not their existence. Always draw them when you have to represent the field between two charged plates.

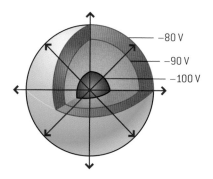

▲ Figure 10.1.3. Field lines and equipotentials around a planet

Figure 10.1.3 shows the gravitational field due to a spherical planet. Points on the green surface are at the same distance from the centre of the sphere and so have the same potential. When a mass moves on the green surface no overall work is done. This gives an *equipotential surface*, on which a charge or mass can move without work being transferred.

Because work is done when a charge or mass moves along a field line, equipotentials must always meet field lines at 90°.

Example 10.1.1

A precipitation system collects dust particles in a chimney. It consists of two large parallel vertical plates, separated by 4.0 m, maintained at potentials of +25 kV and −25 kV.

a) Explain what is meant by an *equipotential surface*.

b) A small dust particle moves vertically up the centre of the chimney, midway between the plates. The charge on the dust particle is + 5.5 nC.

   i) Show that there is an electrostatic force on the particle of about 0.07 mN.

   ii) The mass of the dust particle is 1.2×10^{-4} kg and it moves up the centre of the chimney at a constant vertical speed of 0.80 m s^{-1}.

   Calculate the minimum length of the plates so that the particle strikes one of them. Air resistance is negligible.

Solution

a) An equipotential surface is a surface of constant potential. This means that no work is done in moving charge around on the surface.

b) i) The *force on particle* $= qE = \dfrac{Vq}{d}$ where d is the distance between the plates. The potential difference is 50 kV.

So force $= \dfrac{5.0 \times 10^4 \times 5.5 \times 10^{-9}}{4.0} = 6.875 \times 10^{-5}$ N

  ii) The horizontal acceleration $= \dfrac{\text{force}}{\text{mass}} = \dfrac{6.875 \times 10^{-5}}{1.2 \times 10^{-4}} = 0.573$ m s^{-2}.

The particle is in the centre of the plates, so has to move 2.0 m horizontally to reach a plate. Using $s = ut + \dfrac{1}{2}at^2$ and knowing that the particle has no initial horizontal component of speed gives $2.0 = 0 \times t + \dfrac{1}{2}0.573t^2$ so $t = \sqrt{\dfrac{2 \times 2.0}{0.573}} = 2.63$ s and, therefore, the length must be $2.63 \times 0.8 = 2.1$ m.

>> **Assessment tip**

Example 10.1.1 b) i) is a 'show that' question. You must convince the examiner that you have completed **all** of the steps to carry out the calculation. The way to do this is to quote the final answer to at least one more significant figure (sf) than the question quoted. Here it is quoted to 4 sf—and in this situation this is fine.

SAMPLE STUDENT ANSWER

Explain what is meant by the gravitational potential at the surface of a planet. [2]

This answer could have achieved 2/2 marks:

It is the work done per unit mass to bring a small test mass from a point of infinity (zero PE) to the surface of that planet (in the gravitational field).

▲ There are two marks for this question and two points to make— this answer has them both: work done per unit mass, and the idea of taking the mass (it does not have to be 'small' in a potential definition) from infinity to the surface.

10.2 FIELDS AT WORK

You must know:

✓ the distinction between potential and potential energy

✓ the relationship between potential gradient and field strength

✓ the meaning of escape speed

✓ what is meant by orbital motion, orbital speed and orbital energy

✓ the magnitude of electric field strength and the variation of electric potential inside a charged sphere

✓ aboutforces and inverse-square law behaviour.

You should be able to:

✓ determine the potential energy of a point mass and the potential energy of a point charge

✓ solve problems involving potential energy

✓ determine the potential inside a charged sphere

✓ solve problems involving the orbital speed of a satellite orbiting a planet and the escape speed of an object escaping the gravitational field of a planet

✓ solve problems involving the orbital energy of charged particles in circular orbital motion and masses in circular orbital motion

✓ solve problems involving the forces acting on charges and masses in radial and uniform fields.

In a radial field that obeys inverse-square law behaviour, potential depends on $\frac{1}{r}$, where r is the distance from the origin of the field to the object (for example, a point charge or point mass).

In Topic 10.1 potential was defined to be zero at infinity and is the work done in moving unit mass or unit positive charge from infinity to a particular point. Therefore there is a *potential difference* involved in moving a mass or charge from one (non-infinity) point in a field to another.

> 🔗 This topic builds on Topics 5.1, 6.1, 6.2 and 10.1.

Electric potential in a radial field

$$V_e = -\frac{kq}{r}$$

q is the charge that produces the radial field.

Gravitational potential in a radial field

$$V_g = -\frac{GM}{r}$$

M is the mass of the object producing the radial field.

The potential energy (with zero potential at infinity) in each case is:

$$E_p = qV_e = \frac{kq_1q_2}{r}$$

$$E_p = mV_g = -\frac{GMm}{r}$$

> 🔗 The value of the constant k in the data booklet is $\frac{1}{4\pi\varepsilon_0}$ for a vacuum (or air, approximately), as in Topic 5, and $\frac{1}{4\pi\varepsilon_r\varepsilon_0}$ when in a dielectric material (see Topic 11.3 for more details on the meaning of ε_r).

> 🔗 Electric potential difference links to a term used in Topic 5 – simply called *potential difference* (pd) there.

An important relationship between field strength and potential is

$$\text{field strength} = -\frac{\text{change in potential}}{\text{change in distance}}.$$

In graphical terms, field strength at point x is $-$(gradient of the *potential* against *distance* graph at point x). This is known as the *potential gradient*.

Electric field strength,	Gravitational field strength,
$E = -\dfrac{\Delta V_e}{\Delta r}$	$g = -\dfrac{\Delta V_g}{\Delta r}$

For a uniform electric field, electric field strength is

$$-\frac{(\text{final potential} - \text{initial potential})}{\text{change in distance}};$$

when the change is from $+V$ to 0,

(a change of $-V$), $E = -\left(\dfrac{-V}{d}\right) = \dfrac{V}{d}$.

For a uniform gravitational field close to a planet surface, $g = -\dfrac{\Delta V_G}{h}$, where ΔV_G is the change in potential over a distance h. The change in gravitational potential energy is $m\Delta V_G$, which is mgh, when an object of mass and m moves through vertical distance h.

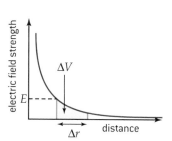

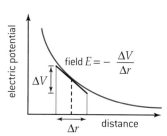

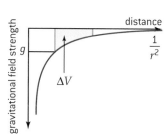

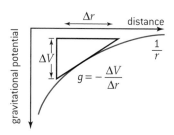

▲ Figure 10.2.1. The relationships between potential and field for both electric and gravitational fields

Example 10.2.1

A charged oil drop of weight 3.0×10^{-14} N is held stationary between two parallel oppositely charged metal plates.

a) The electric field between the plates is uniform.

　i) Explain what this means.

　ii) Sketch field lines to show the electric field between the plates.

b) The plates are separated by 4.0 mm and the potential difference applied between them is 380 V.

Calculate the magnitude of the charge on the oil drop.

Solution

a) i) Field strength is the force per coulomb acting on the oil drop.

　　The field strength is the same everywhere between the plates.

　ii) The field lines should be at 90° to the plates and parallel to each other; the field line separations should be constant. Edge effects should be shown.

b) The field strength $E = \dfrac{V}{d} = 95\,000$ N C^{-1}. The weight of the drop $= Eq$.

So $3.0 \times 10^{-14} = 9.5 \times 10^4 q$ and $q = 3.2 \times 10^{-19}$ C.

Figure 10.2.1 shows, for both electric and gravitational fields, the relationships between field strength and potential. The electric field graph is shown for a positive charge (a repulsive effect). The gravity graph will also apply to attraction in the electric case.

You should understand the distinction between potential and potential energy.

$\left\{\dfrac{\text{electric}}{\text{gravitational}}\right\}$ **potential at a point** is the work per unit $\left\{\dfrac{\text{charge}}{\text{mass}}\right\}$ in

moving from infinity to the point.

$\left\{\dfrac{\text{electric}}{\text{gravitational}}\right\}$ **potential energy** is the work required to move a

$\left\{\dfrac{\text{charge}}{\text{mass}}\right\}$ between two points at different potentials.

Example 10.2.2

A and B are separated by a distance of 200 mm.

Point P is 120 mm from A and 160 mm from B.

Point A has a charge of +2.0 nC.
Point B has a charge of −3.0 nC.

a) Explain why there is a point X on the line AB at which the electric potential is zero.

b) Calculate the distance of the point X from A.

Solution

a) The potential due to A is always positive; the potential due to B is always negative. There must, therefore, be a point (X) at which the potentials sum to zero.

b) Taking x as the distance from A to the zero potential point,

$\dfrac{k \times 2 \times 10^{-19}}{x} + \dfrac{k \times (-3 \times 10^{-19})}{(0.20 - x)} = 0.$ This solves to give $x = 8.0$ cm.

Rockets and satellites that orbit a planet or Sun have both kinetic energy and stored gravitational potential energy when in orbit.

The gravitational potential energy is $E_p = -\dfrac{GM_p m_s}{r}$ where r is the orbital radius (from the centre of the planet and not the surface) M_p is the mass of the planet and m_s is the mass of the satellite.

The kinetic energy of the satellite is $\dfrac{1}{2}m_s v^2$.

Gravity provides the centripetal force required to keep the satellite in its circular orbit. So, $\dfrac{GM_p m_s}{r^2} = \dfrac{m_s v^2}{r} \Rightarrow \dfrac{1}{2}m_s v^2 = \dfrac{GM_p m_s}{2r}$

The term on the left-hand side is the kinetic energy of the satellite and is half the **magnitude** of the gravitational potential energy.

The total energy of the satellite at orbital radius r is $-\dfrac{GM_p m_s}{2r}$.

The equation $\dfrac{GM_p m_s}{r^2} = \dfrac{m_s v^2}{r}$

$\Rightarrow v^2 = \dfrac{GM_p}{r}$

This shows that the **orbital speed**

$v_{orbit} = \sqrt{\dfrac{GM_p}{r}}$.

Notice that v_{orbit} depends only on the orbital radius and the mass of the planet not on the mass of the satellite.

A **geosynchronous orbit** is one with an orbital period that matches the Earth's rotation on its axis.

A special case of this type of orbit is the **geostationary orbit** which is geosynchronous and also positioned above the equator. A satellite in geostationary orbit remains apparently fixed in position when viewed from Earth. This type of orbit is used for many communications satellites.

>> **Assessment tip**

The point that the kinetic energy increases when an orbit decays (radius decreases) is often misunderstood by students.

Remember that $v_{orbit} = \sqrt{\dfrac{GM}{r}}$

(M is the planet mass).

This tells you that if r is smaller, v_{orbit} must be larger.

Escape speed is the minimum speed required for an object to leave the Earth's surface and (just) reach infinity.

The initial kinetic energy must provide energy equal in magnitude to the gravitational potential energy at the surface; this is $\dfrac{GM_p}{r}$.

So $\dfrac{1}{2}m_s v_{esc}^2 = \dfrac{GM_p m_s}{r}$

$\Rightarrow v_{esc} = \sqrt{\dfrac{2GM_p}{r}}$

This is $\sqrt{2} \times$ the orbital speed at the surface.

Equating the gravitational force with the centripetal force leads to other results too. Using the angular velocity ω, rather than the linear speed, gives

$$\frac{GMm}{r^2} = mr\omega^2$$

And, because $T = \dfrac{2\pi}{\omega}$,

$\dfrac{GMm}{r^2} = \dfrac{mr4\pi^2}{T^2}$, which gives

$$T^2 = \frac{4\pi^2}{GM}r^3.$$

The relationship between T and r is due to German mathematician Johannes Kepler; it is known as Kelper's third law. Kepler deduced this empirically from the astronomical observations of Danish astronomer Tycho Brahe.

What happens to a satellite when it moves to a lower orbit? The equations predict that, at a lower radius, the gravitational potential energy is more negative (has a larger magnitude) and the kinetic energy is higher. The total energy is also more negative. The speed of the satellite, therefore, increases even though, overall, it has lost energy.

The energy of a charged particle orbiting another charge of opposite sign follows in a similar way. Equating Coulomb's law with the centripetal force: $\dfrac{kq_1q_2}{r^2} = \dfrac{mv^2}{r} \Rightarrow E_P = 2E_K$, as before.

Rockets can leave Earth's gravity completely as well as remaining bound in orbits. The *escape speed* is the speed required to do this.

Example 10.2.3

The mass of the Earth is 6.0×10^{24} kg and its radius is 6.4×10^7 m.

a) Deduce the radius of a geosynchronous orbit for the Earth.

b) Calculate the change in potential energy of a satellite of mass 750 kg when raised to a geosynchronous orbit from the surface of Earth.

Solution

a) The time period for an Earth geosynchronous orbit is 24 hours.

$T = (24 \times 60 \times 60) = \dfrac{2\pi}{\omega}$. Therefore, $\omega = 7.3 \times 10^{-5}$ rad s^{-1}

Equating centripetal force with gravitational force leads to

$$r = \left(\frac{GM}{\omega^2}\right)^{\frac{1}{3}}.$$

Substituting gives an answer for the orbital radius of 4.2×10^7 m.

b) The change in potential $= -GM\left(\dfrac{1}{r} - \dfrac{1}{R}\right)$ where R is the Earth radius and r is the orbital radius.

This is $6.67 \times 10^{-11} \times 6.0 \times 10^{24}\left(\dfrac{1}{6.4 \times 10^6} - \dfrac{1}{4.2 \times 10^7}\right)$

$= 5.3 \times 10^7$ J kg^{-1}.

(Note the manipulation of the negative signs here – never ignore the signs; always carry them through.)

The potential energy change is $\Delta V_g \times m_{satellite} = 5.3 \times 10^7 \times 750$
$= 4.0 \times 10^{10}$ J

Moving charged particles are affected by the presence of electric and magnetic fields. The fields give rise to different types of motion for the particles.

Charged particles in an electric field

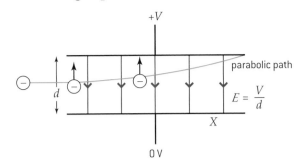

Charged particles in a magnetic field

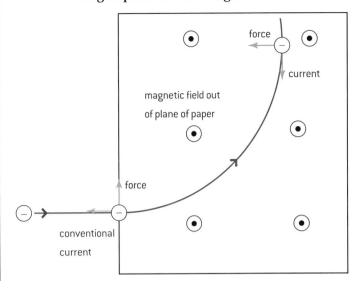

An electron enters a uniform electric field perpendicular to the field lines. A force acts in the opposite direction to these lines.

The force on the charge is Ee and the acceleration $a = \dfrac{Ee}{m_e}$. In terms of the pd V between the charged plates that produce the field and their separation d: $a = \dfrac{eV}{m_e d}$.

An electron enters a uniform magnetic field perpendicular to the field lines. The magnetic force acts *always* at 90° to the velocity of the electron. This is the condition for a centripetal force and the electron moves in a circular path.

The force on the electron is Bev which must equal $\dfrac{m_e v^2}{r}$, where r is the radius of the orbit.

Example 10.2.4

A proton and a positive pion travel along the same path at a speed of $1.5 \times 10^7\,\text{m s}^{-1}$ into a uniform magnetic field of flux density 0.16 T. The magnetic field direction is at 90° to the initial direction of motion of the proton and the pion. The rest mass of a pion is $2.5 \times 10^{-28}\,\text{kg}$.

a) Calculate the radius of curvature of the path of the proton.

b) Comment on how the path of the pion will differ from that of the proton.

c) The magnetic field strength is decreased. Suggest how this affects the paths of the particles.

Solution

a) Rearranging and simplifying $\dfrac{m_p v^2}{r} = Bev$ gives $r = \dfrac{m_p v}{eB}$.

The data booklet provides m_p; substituting gives

$$r = \frac{1.67 \times 10^{-27} \times 1.5 \times 10^7}{1.6 \times 10^{-19} \times 0.16} = 0.98\,\text{m}.$$

b) A pion has less mass than a proton. The particles have the same charge so only the mass is different. From the radius equation, r is smaller for the pion and so its path is more curved.

c) Again, from the radius equation, when B is decreased, r increases and so the path of both particles is less curved.

The gravitational potential due to the Sun at its surface is $-1.9 \times 10^{11}\,\text{J kg}^{-1}$. The following data are available.

Mass of Earth	$= 6.0 \times 10^{24}\,\text{kg}$
Distance from Earth to Sun	$= 1.5 \times 10^{11}\,\text{m}$
Radius of Sun	$= 7.0 \times 10^{8}\,\text{m}$

a) Outline why the gravitational potential is negative. [2]

This answer could have achieved 0/2 marks:

It is works in the opposite direction to the gravitational force.

▼ While there is some truth in the answer, it does not get us very far. This answer is really answering a question about the relationship between gravitational force and the defined positive direction.

This answer could have achieved 2/2 marks:

Because 0 of gravitational potential is defined at infinity. Since gravity attracts, work always has to be done on the object to get closer to infinity, which is the same thing as closer to 0 potential, so potential is negative, since work need to be added to move away from the Sun's gravitational field.

▲ There are two marks and therefore two points need to be made. This answer scores both: (i) potential is zero at infinity, (ii) work has to be done to move a mass away from the Sun to infinity. The potential before the mass was moved must have been negative for it to have gained potential and still be zero at infinity.

b) The gravitational potential due to the Sun at a distance r from its centre is V_s. [1]

Show that

$$rV_s = \text{constant}$$

This answer could have achieved 0/1 marks:

▼ The first sentence is not very clear. It should be expressed the other way round for clarity. Fortunately the remainder of the answer makes up for this.

$$V_s = \frac{GM}{r} \quad rV_s = -GM$$

Mass of the Sun

▼ You maybe surprised that this answer scored zero. The key is in the command term: *show that*. The answer correctly rearranges the equation for gravitational potential and shows that the value of the constant is GM, but that is not enough. It should have gone on to say that G is constant. Therefore, $G \times M$ must be constant too.

>> **Assessment tip**

This may seem picky to you but it is the essence of a *show that* question. You must not expect the examiner to finish the work for you (which is what this answer requires)—write every step down.

Practice problems for Topic 10

Problem 1

Two parallel metal plates A and B are fixed vertically 20 mm apart and have a potential difference of 1.5 kV between them.

a) Sketch a graph showing the potential at different points in the space between the plates.

b) A plastic ball of mass 0.5 g is suspended midway between the plates by a long insulating thread. The ball has a conducting surface and carries a charge of 3.0 nC.

   The ball is released from rest. Deduce the subsequent motion of the ball.

Problem 2

a) Outline what is meant by gravitational field strength.

b) The radius of Jupiter is 7.1×10^4 km and the gravitational field strength at the surface of Jupiter is 25 N kg^{-1}. Estimate the mass of Jupiter.

Problem 3

A proton from the Sun passes above the North magnetic pole moving parallel to the Earth's surface at a speed of 1.2×10^6 m s^{-1}. At this point the magnetic field of the Earth is vertically downwards with a magnetic flux density of 5.8×10^{-5} T.

a) Calculate the radius of the path of the proton when it is above the pole.

b) A helium nucleus moving with the same initial velocity of the proton is also above the pole.

   Compare the paths of the proton and the helium nucleus.

Problem 4

a) Define electric potential at a point.

b) A metal sphere of radius 0.060 m is charged to a potential of 450 V.

   (i) Deduce the magnitude of the electric charge on the sphere.

   (ii) Determine the magnitude of the electric field strength at a distance of 0.12 m from the centre of the sphere. State an appropriate unit for your answer.

   (iii) Identify the magnitude of the gradient of electric potential at a distance of 0.12 m from the centre of the sphere.

Problem 5

A satellite orbiting a planet has an orbital period of 460 minutes and an orbital radius of 9.4 Mm.

a) The satellite orbits with uniform circular motion. Outline how this motion arises.

b) Show that the orbital speed of the satellite is about 2 km s^{-1}.

c) Deduce the mass of the planet.

11 ELECTROMAGNETIC INDUCTION (AHL)

11.1 ELECTROMAGNETIC INDUCTION

You must know:

✔ that there is a magnetic flux in a region where there is a magnetic field

✔ what is meant by magnetic flux linkage

✔ that magnetic flux density is equivalent to magnetic field strength

✔ that an electromotive force (emf) is induced in a conductor when it moves relative to a magnetic field or when there are changes in the magnetic flux that links the conductor

✔ Lenz's law and Faraday's law of induction.

You should be able to:

✔ describe how the induced emf arises when a magnetic flux changes and when a conductor moves relative to a uniform magnetic field

✔ solve problems that include the use of Faraday's law of induction and magnetic flux and magnetic flux linkage changes

✔ explain how conservation of energy leads to Lenz's law

✔ explain that the magnitude of the induced emf depends on the change in magnetic flux linkage.

🔗 Topic 5.4 showed that electric currents in conductors in magnetic fields lead to motion. Magnetic fields were visualized using field lines. The closer these were drawn, the stronger the field represented. The concept of magnetic field strength (symbol B was used in Topic 5.4) is developed further.

🔗 This idea links to Topic 5.4. There, charge in a wire was moving in a magnetic field and this gave rise to a force acting on a wire. Here the wire is moving its electrons along with it – but the physics is the same.

Scientists in the 1820s realised that relative motion of a conductor in a magnetic field leads to the production of an induced emf and to an induced current in the conductor. This introduces the concepts of *magnetic flux, magnetic flux density* and *magnetic flux linkage*.

Magnetic flux ϕ is related to the number of lines cut by a conductor or enclosed by a one-turn coil while in the magnetic field. The unit of magnetic flux is the weber (Wb).

The density of field lines indicates the strength of the field, so the magnetic field strength, measured in tesla (T), can also be called the **magnetic flux density** (in symbols, $\phi = BA$). Density in this case means the number of field lines per square metre.

The units of magnetic flux density are $Wb\,m^{-2}$.

$1\,T \equiv 1\,Wb\,m^{-2}$.

When a coil has more than one turn (N turns), then each coil turn links one set of lines, and the total **magnetic flux linkage** is (flux for one turn) × (number of turns); in other words, NBA or $N\phi$.

An emf is generated between the ends of a straight conductor when it moves at right angles to a magnetic field. In Figure 11.1.1 the magnetic field is into the plane of the page, and the conducting rod is moving in the plane of the page.

magnetic field into page

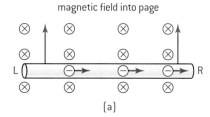

(a)

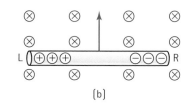

(b)

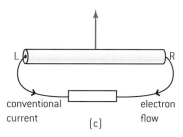

conventional current

electron flow

(c)

▲ Figure 11.1.1. An emf is induced in a conductor moving in a magnetic field

Individual free electrons are moved upwards by the moving rod. This is a conventional current moving downwards and a motor effect acts on each electron. Fleming's left-hand rule (or other rule) gives the force on each free electron to the right. This produces a negative charge at R and an electron deficit (positive) at L. Eventually, the electron flow stops as electric forces prevent further motion along the rod. An *induced electromotive force* (emf) has been generated in the rod.

When there is a complete circuit, as shown in Figure 11.1.1(c), charge will flow. Kinetic energy is transferred to electrical energy. *Lenz's law* predicts that this current will produce a motor effect to the left to oppose the change made to the system.

Work is done to overcome this opposing force and is the origin of the transferred energy in the system. Lenz's law is equivalent to conservation of energy. If the forces produced acted upwards, energy would be created in the system. This is not possible.

Faraday's law makes a quantitative statement about the magnitude of the electromagnetic effect. In this case, when the conductor of length *l* moves at speed *v* in a uniform magnetic field of flux density *B*, the change in the flux every second is Bvl because $\varepsilon = -NB \times \dfrac{lv}{1}$ and $N = 1$.

When there are N turns, $\varepsilon = BvlN$.

Faraday and others developed the idea that the conductor in Figure 11.1.1 was 'cutting' field lines; this is a simple idea to grasp. It can be extended to other conductor shapes too.

When the magnetic field moves and the conductor remains fixed, or when the magnetic field changes magnitude and the conductor remains fixed, the effects are the same as when the conductor moves in the magnetic field.

The magnetic flux density changes by ΔB over a time Δt. As a result, Faraday's law predicts that the emf generated will be $-NA\dfrac{\Delta B}{\Delta t}$, where N is the number of turns, A is the fixed area of the coil and $\dfrac{\Delta B}{\Delta t}$ is the rate at which the magnetic flux density changes with time.

Lenz's law states that the induced emf gives rise to an induced current that opposes the change producing it. It provides a statement about the direction of the electromagnetic induction effect.

Faraday's law states that the emf ε induced is given by $\varepsilon = -N\dfrac{\Delta\phi}{\Delta t}$, where $\Delta\phi$ is the change in flux and Δt is the change in time. N is the number of turns of the coil for cases where this applies. When the velocity and field are not perpendicular, $\Phi = BA\cos\theta$. This can also be written as $\varepsilon = -N\dfrac{\Delta(BA\cos\theta)}{\Delta t}$, where B and A are the magnetic flux density and the area swept out respectively, and θ is the angle between the field and the conductor.

See Topic 5.1 for a discussion more about the meaning of conventional current.

Example 11.1.1

A coil with five turns has an area of 0.25 m². The magnetic flux density in the coil changes from 60 mT to 30 mT is a time of 0.50 s.

Calculate the magnitude of the emf induced in the coil.

Solution

$$\varepsilon = -NA\frac{\Delta B}{\Delta t}$$

The magnitude is

$$-5 \times 0.25 \times \frac{(60-30) \times 10^{-3}}{0.50}$$

$$= 75\,\text{mV}$$

SAMPLE STUDENT ANSWER

A vertical metal rod of length 0.25 m moves in a horizontal circle about a vertical axis in a uniform horizontal magnetic field.

The metal rod completes one circle of radius 0.060 m in 0.020 s in the magnetic field of strength 61 mT.

Determine the maximum emf induced between the ends of the metal rod. [3]

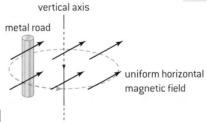

vertical axis
metal road
uniform horizontal magnetic field

This answer could have achieved 2/3 marks:

Speed of rotation $= 2\pi \times 0.06 \div 0.02 = 0.6\pi\,\text{m s}^{-1}$

maximum emf is $Bvl = 61 \times 10^{-3} \times 0.25 \times 0.6\pi$

$= 9.15\,\text{mV}$

▼ This attempt has one error. The calculation of the rotation speed is incorrect (it should be 6.0π m s⁻¹). The final answer would have been better to two significant figures.

▲ The solution is well explained and the steps are clear.

11.2 POWER GENERATION AND TRANSMISSION

You must know:

✔ what is meant by an alternating current generator

✔ the meaning of average power and root mean square values for current and voltage

✔ what is meant by a step-up and a step-down transformer

✔ what is meant by a diode bridge

✔ what is meant by full-wave and half-wave rectification

✔ how to describe qualitatively the effect of adding capacitance to a diode-bridge rectification circuit.

You should be able to:

✔ explain the operation of an alternative current (ac) generator including changes of generator frequency

✔ solve problems involving average power and problems involving root mean square (rms) and peak currents and voltages in ac circuits

✔ describe the use of transformers in the transmission of ac electrical energy

✔ investigate diode bridge rectification circuits experimentally

✔ solve problems involving step-up and step-down transformers.

> 🔗 The principles of electromagnetic induction discussed in Topic 11.1 apply to coils rotating in a magnetic field that lead to the generation of alternating currents (ac).

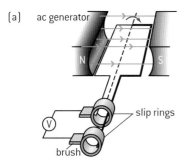

(a) ac generator — slip rings — brush

(b)

▲ Figure 11.2.1. Simple ac generator

The simplest ac generator is a one-turn coil of wire rotating in a uniform magnetic field (Figure 11.2.1(a)). Zero flux links the coil (Figure 11.2.1(b)) when the coil is parallel to the field lines. When the coil has turned through 90° it is perpendicular to the field lines and the flux has changed, inducing an emf. When the coil turns through another 90° it is in its original orientation except that the sides of the coil have reversed. The flux linkage has changed again. The coil rotates through a final 180° to return to its original orientation.

The emf in the coil varies continuously between a positive peak and a negative trough (Figure 11.2.2). With suitable electrical connections, an alternating current will be induced in a load connected to the coil.

Slip rings (Figure 11.2.1(a)) are used to transfer energy from the coil to the load. The rings are attached to the coil ends and rotate with it. Stationary brushes, often made of conductive carbon, press onto the rings and are connected to the load.

Example 11.2.1

A rectangular coil of 650 turns with dimensions 20 mm × 35 mm rotates about a horizontal axis. The axis is at right angles to a uniform magnetic field of flux density 2.5 mT. At one instant the plane of the coil makes an angle θ with the vertical.

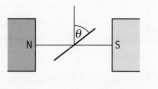

a) Identify the value of θ for the magnitude of the magnetic flux through the coil to be a minimum.

b) Calculate the magnetic flux passing through the coil when θ is 30°.

c) Determine the maximum flux linkage through the coil as it rotates.

Solution

a) The coil must be parallel to the field for the flux to be zero (this is the minimum magnitude). θ is either 90° or 270°.

b) The magnetic flux is $\phi = BA \cos\theta = 1.5\,\mu\text{Wb}$.

c) The maximum flux linkage is when the coil is perpendicular to the field ($\theta = 0$) which corresponds to a maximum flux of 1.75 μWb. The maximum flux linkage is: $650 \times 1.75 \times 10^{-6} = 1.1\,\text{mWb turns}$.

Figure 11.2.2 shows the output of an ac generator that is rotating at a constant speed; the waveform is sinusoidal. Quantities associated with the varying output are its frequency and the peak emf ε_0 and peak current I_0.

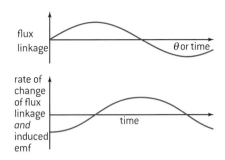

▲ **Figure 11.2.2.** Variation of emf with time for one cycle of a coil in a uniform magnetic field

Alternating current and pd values can be compared with direct current (dc) values. A suitable average is needed for ac. When the current and pd are averaged over one cycle, their mean value is zero and cannot be used. However, the power P dissipated in the load is always positive.

Figure 11.2.3 shows that the wave shape of the power dissipated is always positive ($\sin^2$) and has a non-zero average that is equal to $\dfrac{V_0 \times I_0}{2}$. This can be written as $\dfrac{V_0}{\sqrt{2}} \times \dfrac{I_0}{\sqrt{2}}$.

These separate quantities are known as the root-mean-square (rms) values of the current and pd, with $V_{rms} = \dfrac{V_0}{\sqrt{2}}$ and $I_{rms} = \dfrac{I_0}{\sqrt{2}}$.

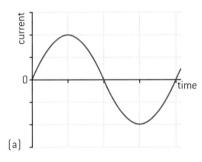
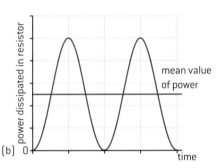

▲ **Figure 11.2.3.** Current, power and mean power values for ac

The power delivered over one cycle by an alternating supply to a resistor of resistance R is $V_{rms}I_{rms}$. Root mean square values are the *effective* values of an alternating power supply. A lamp connected to a 6 V direct current supply has the same brightness when connected to a 6 V rms alternating supply.

Example 11.2.2

An ac power supply of emf V_0 and negligible internal resistance is connected to two identical resistors of resistance R in parallel.

Calculate the average power dissipated in the circuit.

Solution
The total resistance in the circuit is $\dfrac{R}{2}$ as the resistors are in parallel.

The power dissipated in the circuit is $\dfrac{2V_{rms}^2}{R}$.

However, $V_{rms} = \dfrac{V_0}{\sqrt{2}}$, so the power dissipated is $\dfrac{V_0^2}{R}$.

Alternating current can easily be converted from one voltage to another using a transformer. This is useful in domestic situations (to connect low-voltage devices to the mains) or to transmit electrical energy over long distances.

>> **Assessment tip**

When the rotation speed is doubled:

• the frequency doubles because the time for one rotation halves

• the rate of change of flux linkage doubles and so the peak emf doubles too.

>> **Assessment tip**

The resistance of a device in an alternating circuit can be calculated using either

$R = \dfrac{V_0}{I_0}$ or $\dfrac{V_{rms}}{I_{rms}}$. However, you must use either rms or peak values – do not mix them.

The peak power delivered by the supply is $P_{max} = V_0 I_0$.

This is only delivered twice in each cycle at the instant of each maximum and minimum of the waveform.

The average power is $V_{rms} I_{rms}$ which is $\dfrac{1}{2}V_0 I_0$.

The input voltage V_s and output emf ε_s are related to the ratio of the numbers of turns: $\dfrac{V_p}{\varepsilon_s} = \dfrac{n_p}{n_s}$ where n_p is the number of turns on the primary coil and n_s is the number of turns on the secondary.

This is written in the data booklet as $\dfrac{\varepsilon_p}{\varepsilon_s} = \dfrac{N_p}{N_s} = \dfrac{I_p}{I_s}$.

Energy is conserved in the transformer (this is approximately true in practical transformers where efficiencies are 90% or higher) so $V_p I_p = \varepsilon_s I_s$, where I_p and I_s are the primary and secondary currents. In other words, the power input to the device is equal to the power output.

Energy losses in the transformer

Electrical resistance in the coils leads to I^2R losses that can be reduced by using thick wires for coils and connecting wires.

Core losses include the following.

• **Eddy current losses**: The changing magnetic field produces an induced emf in the conducting core as well as in the secondary. This loss is reduced by laminating the core – manufacturing it in insulated strips so that the electrical resistance is large but the magnetic properties are unaffected.

• **Hysteresis losses**: Energy is needed to re-magnetize the core in opposite directions.

Less than 100% of the core flux will be linked to the secondary leading to inefficiency.

Transformers consist of:

• a primary coil connected to the input terminals

• a secondary coil connected to the output terminals

• a core made from a soft iron material that is easily magnetised in the presence of a magnetic field but loses this magnetism quickly when the field is turned off.

Both coils are wound on the core so that the fields through both are as similar as possible.

The operation of the transformer

• Alternating current is supplied to the primary and establishes an alternating magnetic field in the core. This field reverses direction at the same rate as the alternating current.

• As both coils are wound on the same core, this field is linked to the secondary.

• There is a constantly changing magnetic field inside the secondary and an emf is induced in it.

• When there is a resistive load attached to the secondary, there is an induced current in the load.

Example 11.2.3

An ideal transformer with 800 turns produces an rms output of 2.0×10^3 V when it is connected to the 230 V rms mains supply.

a) Calculate the number of turns required on the secondary coil.

b) Outline how *eddy currents* are minimized in a transformer.

Solution

a) $\dfrac{V_P}{\varepsilon_s} = \dfrac{n_P}{n_s}$; $n_s = 6960$ turns

b) There is an induced emf in the core due to the changing magnetic field in it. This leads to an induced current because the core is metallic and therefore a conductor. By using laminations, the core has a high electrical resistance but is still effective in allowing magnetic flux to penetrate it.

The conversion of one ac voltage to another is particularly important for energy transmission over long distances; it allows a reduction in energy loss in the cables.

Example 11.2.4

A cable is to transmit 100 kW of electrical power from a power station to a factory. The total resistance in the cable is 0.40 Ω.

Compare the power losses for a transmission voltage of

a) 250 V   b) 25 kV.

Solution

a) 250 V ac: $P = VI$ so the current in the cable is 400 A. The power loss in the cable is $I^2R = 64\,000$ W—a substantial fraction of the total power transmitted.

b) 25 kV ac: The current in the cable is now 4.0 A and the power loss is 6.4 W. There will be significant savings in power loss with the higher pd.

Some devices cannot use alternating current—they require direct current.

The *half-wave rectifier* produces a pulse of current for half of every ac cycle.

The diode allows conduction only in one direction (Figure 11.2.4) so it conducts for only half of each cycle. The input and output waveforms are shown. Note that the average for the output current is not half the peak values.

A full-wave rectified output with two pulses per ac cycle can be obtained in two ways:

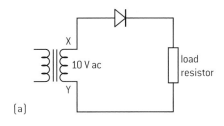

(a)

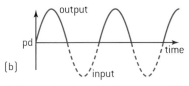

(b)

▲ Figure 11.2.4. Half-wave rectifier

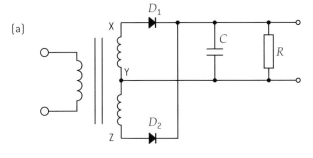

(a)

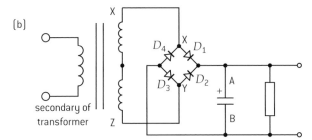

(b)

▲ Figure 11.2.5. Full-wave rectification

> A **rectifier** can be used to convert ac to dc.
>
> A **diode bridge** is one kind of rectifier. Two types are common: the half-wave rectifier and the full-wave rectifier.

Figure 11.2.5(a) shows a two-diode arrangement. When the transformer output is positive at the top of the diagram and negative at the bottom, then diode D_1 conducts. When the transformer output has reversed and the positive output is at the bottom of the secondary coil, then diode D_2 conducts. Whichever diode conducts, however, the current in the load resistor is always in the same direction.

The alternative bridge arrangement (Figure 11.2.5(b)) uses four diodes. When X is positive with Z negative, diodes D_1 and D_3 conduct making point A positive and point B negative. When point Z is positive, diodes D_2 and D_4 conduct giving the same current directions as before.

One way to smooth out the waveform is to place a capacitor in parallel with the load resistor (Figure 11.2.5(a) and (b)).

When the output goes positive for the first time, the capacitor charges. As the output drops towards the zero value, the capacitor discharges but at a slow rate because the value of RC is chosen to be long compared with the time for half a cycle. The voltage, therefore, drops only slightly before the capacitor is charged for the second time.

> The basic properties of diodes are covered in Topic 5.2 where the V–I characteristic is given. When the diode is forward biased, charge can flow through it. But if the pd across the device is reversed, it cannot conduct. This is an ideal way to strip off the negative part of an ac waveform and just leave the positive-going part.

> The charging action of capacitors is discussed in Topic 11.3.

In an alternating current (ac) generator, a square coil ABCD rotates in a magnetic field.

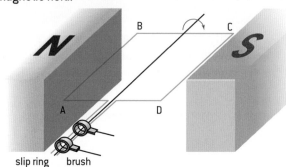

slip ring   brush

The ends of the coil are connected to slip rings and brushes. The plane of the coil is shown at the instant when it is parallel to the magnetic field. Only one coil is shown for clarity.

Explain, with reference to the diagram, how the rotation of the generator produces an electromotive force (emf) between the brushes.   [3]

This answer could have achieved 1/3 marks:

▼ A good answer to this question requires references to:

• the concept of flux *linkage*

• Faraday's law and the rotation of the coil

The emphasis in this answer on the direction of forces and currents is not required by the question.

> The rotation of the square coil changes its flux over time and according to Faraday's law, an emf will be produced as emf is the rate of change of flux. In the diagram, as AB rotate upwards, experiencing a change in flux in the magnetic field between two ends of the magnet, the force is upwards, the field direction is from N→S (rightwards). Using Fleming's left right rule, the current will flow from A to B.

11.3 CAPACITANCE

You must know:

✓ what is meant by capacitance

✓ what is meant by a dielectric material

✓ how capacitors combine in series and parallel

✓ the properties of a resistor–capacitor (RC) circuit

✓ what is meant by the time constant of an RC circuit

✓ how to solve problems involving the discharge of a capacitor through a fixed resistor

✓ how to solve problems for charge, current and voltage involving the time constant of an RC circuit.

You should be able to:

✓ solve problems involving parallel-plate capacitors

✓ describe the effect of dielectric materials on capacitance

✓ investigate combinations of series or parallel capacitor circuits

✓ describe the charging process for a capacitor in series with a resistor

✓ determine the energy stored in a charged capacitor

✓ describe the nature of exponential discharge of a capacitor through a resistor.

A simple capacitor consists of two parallel metal plates, separated by air or an insulator (Figure 11.3.1(a)).

When the capacitor is connected to a power supply, charge flows onto the plates (Figure 11.3.1(b)). This flow eventually ceases when the potential difference between the plates equals that of the power supply (Figure 11.3.1(c)). There is now no longer sufficient energy to transfer more electrons to and from the plates and the capacitor is said to be *fully charged*.

Figure 11.3.2 shows how the pd between the plates varies with charge stored on one of them.

This linear graph leads to the definition of *capacitance*.

Capacitance: $C = \dfrac{q}{V}$, where q is the charge stored and V is the potential difference at which it is stored.

The unit is the farad (F); $1\,\text{F} \equiv 1\,\text{C}\,\text{V}^{-1}$. In fundamental units, this is $\text{m}^{-2}\,\text{kg}^{-1}\,\text{A}^2\,\text{s}^4$.

The energy stored at pd V by a capacitor of capacitance C is $\dfrac{1}{2}qV$ (this is the area under the graph) where q is the charge stored.

The definition of C as $\dfrac{q}{V}$ gives two variants: $\dfrac{1}{2}qV = \dfrac{1}{2}CV^2 = \dfrac{1}{2}\dfrac{q^2}{C}$.

Energy is transferred to a charging capacitor as the pd across it increases. The area under the graph (Figure 11.3.2) gives the energy transferred during charging.

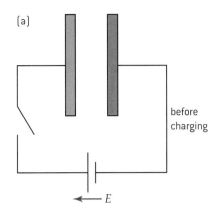

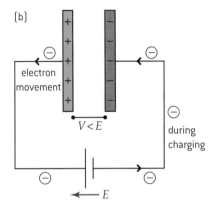

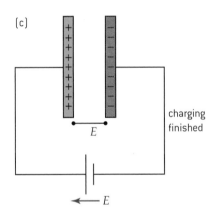

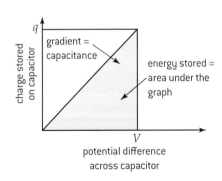

▲ Figure 11.3.1. Charging a capacitor

Example 11.3.1

A charge of $20\,\mu\text{C}$ is stored at a potential difference of $10\,\text{V}$. Calculate the energy stored.

Solution

$$E = \frac{1}{2}qV = \frac{20 \times 10^{-6} \times 10}{2} = 1.0 \times 10^{-4}\,\text{J}$$

qV of energy is transferred from a power supply but only $\dfrac{1}{2}qV$ of energy is stored by the capacitor during charging. Where is the remainder? It has been lost in the resistance of the wires that connect the power supply and capacitor and in the internal resistance of the power supply. When the resistances are very small, the bulk of the energy is radiated away from the system as electromagnetic radiation.

Example 11.3.2

A capacitor stores $20\,\mu\text{C}$ of charge at a potential difference of $10\,\text{V}$. Calculate the capacitance of the capacitor.

Solution

$$C = \frac{q}{V} = \frac{20 \times 10^{-6}}{10} = 2.0 \times 10^{-6} = 2.0\,\mu\text{F}$$

charge stored on capacitor

gradient = capacitance

energy stored = area under the graph

V potential difference across capacitor

▲ Figure 11.3.2. Charge–pd graph for a $470\,\mu\text{F}$ capacitor

Capacitors consist of two plates of area A separated by distance d. Capacitance is given by $C = \varepsilon \dfrac{A}{d} = \varepsilon_0 \varepsilon_r \dfrac{A}{d}$, where the constant ε_r is the relative permittivity of the material between the plates. ε_r for air is very close to 1; however, for other dielectrics, ε_r takes values that range from 1 to 10^6. A dielectric between the plates *increases* the capacitance so that a *greater* charge can be stored at a particular pd.

Example 11.3.3

Two capacitor plates are separated by a rubber sheet of thickness 3.0 mm that fills the space between them. The plate dimensions are 0.35 m × 0.45 m. ε_r for the rubber is 6.0.

Calculate the capacitance of the arrangement.

Solution

The area of the plates is 0.158 m². The distance apart is 3×10^{-3} m.

$$C = \varepsilon_0 \varepsilon_r \frac{A}{d} \text{ gives } C = 8.85 \times 10^{-12} \times 6 \times \frac{0.158}{3 \times 10^{-3}} = 2.8 \times 10^{-9} \, \text{F} \equiv 3 \text{ nF}$$

Dielectric materials are polar, meaning that the insulator molecule has one end positive and the other negative, either permanently or temporarily (in an electric field).

The molecules line up in the field with the positive ends towards the negative plate. This reduces the total field: $E_{net} = E_{cap} - E_{dielectric}$. Because the total field $E_{net} = \dfrac{V}{d}$, V is also reduced. The dielectric *increases* capacitance.

Capacitors can be combined in series or parallel networks.

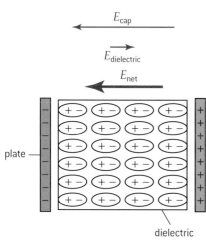

▲ Figure 11.3.3. How a dielectric increases capacitance

Capacitors in parallel

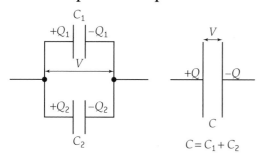

$$C = C_1 + C_2$$

The same pd V must be across each capacitor.

The total charge stored (and supplied by the power supply) is $Q = Q_1 + Q_2$.

This can be written as

$CV = C_1V + C_2V$ and $C_{parallel} = C_1 + C_2 + \cdots$.

Capacitors in series

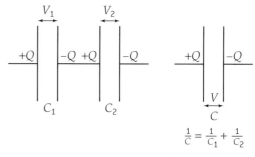

$$\frac{1}{C} = \frac{1}{C_1} + \frac{1}{C_2}$$

A charge Q flows onto each capacitor – as they are in series, this charge is the same for both.

The pds are given by $V = \dfrac{Q}{C}$ in both cases.

So $V = V_1 + V_2 = \dfrac{Q}{C_1} + \dfrac{Q}{C_2} = \dfrac{Q}{C}$, where C is the capacitor equivalent to 1 and 2 in series.

Thus, $\dfrac{1}{C_{series}} = \dfrac{1}{C_1} + \dfrac{1}{C_2} + \cdots$

Example 11.3.4

A 20 μF and a 60 μF capacitor are connected in series. This series arrangement is then connected in parallel with capacitor X. The capacitors are fully charged.

The potential difference across X is 12 V and the total charge stored is 4.0×10^{-4} C.

a) Calculate the total capacitance of the three capacitors.

b) Calculate the capacitance of X.

c) Calculate the potential difference between the terminals of the 60 μF capacitor.

Solution

a) $C = \dfrac{q}{V}$ $C = \dfrac{4.0 \times 10^{-4}}{12} = 33 \ \mu F$

b) For the series capacitors, $\dfrac{1}{C} = \dfrac{1}{20 \times 10^{-6}} + \dfrac{1}{60 \times 10^{-6}}$. C is 15 μF.

Parallel capacitances add, therefore X must have a capacitance of $(33.3 - 15) = 18 \ \mu F$.

c) The same charge is stored on both series capacitors and $V = \dfrac{q}{C}$.

So, the pd must be split in inverse ratio to the capacitances. The ratio is 1 : 3 so the pd across the 60 μF capacitor is 3.0 V.

When a capacitor loses its charge through a fixed resistor, the change is *exponential*.

Figure 11.3.4 shows the variation in charge Q with time t for a discharging capacitor.

The decay constant λ in radioactive decay is analogous to $\dfrac{1}{\tau}$. There is, however, no formal concept of half-life in capacitance theory even though there is still a constant time for the initial charge to halve. Instead, remember that the charge reduces to 37% of its original value in time τ (Figure 11.3.5).

Because $C = \dfrac{q}{V}$, when C is constant, $q \propto V$, so just re-scale the axes to move from a Q–t to a V–t graph.

The current during discharge is $I = I_0 e^{\left(-\frac{t}{\tau}\right)}$.

Example 11.3.5

A 470 μF capacitor discharges through a resistor of resistance R. The initial pd at time $t = 0$ is 12 V and at $t = 1.0$ s the pd across the resistor is 8.0 V.

Calculate R.

Solution

pd across the capacitor at $t = 1.0$ s is $(12 - 8) = 4.0$ V so $4 = 12 e^{-\frac{1}{RC}}$

Rearranging and taking logs: $\ln\left(\dfrac{1}{3}\right) = -\dfrac{1}{RC}$ and $\ln(3) = 1.10 = \dfrac{1}{RC}$.

$R = \dfrac{1}{1.10 \times 4.7 \times 10^{-4}} = 1940 \equiv 19 \ k\Omega$

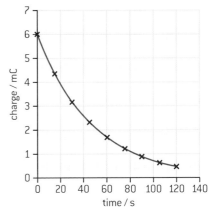

▲ **Figure 11.3.4.** Charge versus time for capacitor discharge

The **discharge equations** are

$q = q_0 e^{\left(-\frac{t}{\tau}\right)}$ and $V = V_0 e^{\left(-\frac{t}{\tau}\right)}$,

where q_0 and V_0 are the initial charge and pd, q and V are the charge and pd remaining at time t and τ is the time constant which determines the discharge rate.

$\tau = RC$

Also, since current I is the rate of change of charge, $I = I_0 e^{\left(-\frac{t}{\tau}\right)}$.

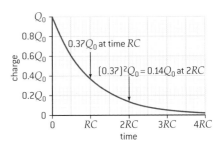

▲ **Figure 11.3.5.** Discharge graph in units of RC

≫ Assessment tip

There are close similarities between the way the stored charge changes with time and the way the number of nuclei decreases with time in radioactive decay. Use these similarities to aid your understanding.

The mathematics of radioactive decay is covered in Topic 12.2; the physics of half-life is discussed in Topic 7.1.

When a capacitor charges through a fixed resistance, the charging curve is asymptotic to the final q and V values. So, in principle, this takes an infinite time. In practice, the q and V values reach 99% by a time $5RC$.

SAMPLE STUDENT ANSWER

An uncharged capacitor in a vacuum is connected to a cell of emf 12 V and negligible internal resistance. A resistor of resistance R is also connected.

At $t = 0$ the switch is placed at position A. The graph shows the variation with time t of the voltage V across the capacitor. The capacitor has capacitance 4.5 μF in a vacuum.

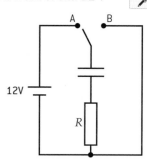

a) On the axes, draw a graph to show the variation with time of the voltage across the resistor.   [2]

This answer could have achieved 0/2 marks:

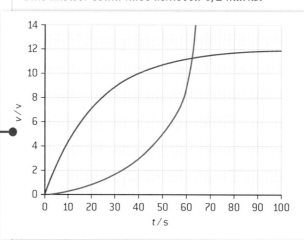

▼ There are misunderstandings about the functions of the components. The capacitor and R are in series; the total pd across them both must always be 12 V. When the pd across the capacitor is 2 V, the pd across R must be 10 V and so on. The two curves will cross at 6 V. The voltage for the resistor will be asymptotic to the x-axis as the capacitor pd is asymptotic to 12 V.

b) i) The time constant of this circuit is 22 s. State what is meant by the time constant.   [1]

This answer could have achieved 0/1 marks:

The time constant is the time it takes to charge up half of the capacitor.

▼ The time constant is RC and it has the units of s when R is in ohms and C is in farads. It is also the time taken for the initial charge on a discharging capacitor to fall to 37% of its initial value.

ii) Calculate the resistance R.   [1]

This answer could have achieved 0/1 marks:

▼ The answer ignores the factor of 10^{-6} from the 4.5 μF and with only one mark for the question, all credit is lost.

$$T = R.C \qquad \frac{T}{C} = R \qquad T = \text{time constant} \qquad C = \text{capacitance}$$

$$R = \text{resistance} \quad \frac{22s}{4.5 \mu F} = 4.9\Omega$$

c) A dielectric material is now inserted between the plates of the fully charged capacitor. State the effect, if any, on

i) the potential difference across the capacitor.   [1]

This answer could have achieved 0/1 marks:

▼ The pd across the RC combination is determined by the power supply while the switch is at position A (we are not told that it is moved). Even though there is charge movement, the pd stays the same.

The potential difference will increase.

ii) the charge on one of the capacitor plates.   [1]

This answer could have achieved 1/1 marks:

▲ This is correct although it is rather too brief. A dielectric material is introduced between the plates, increasing the capacitance. The pd is constant (maintained by the power supply), so the charge stored must have increased too.

The charge on one of the plates will increase.

Practice problems for Topic 11

Problem 1

The graph shows the variation of magnetic flux ϕ through a coil with time t.

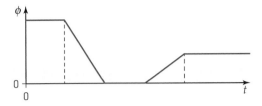

a) Sketch the variation of the magnitude of the emf in the coil with time over the same time period.

b) Explain your answer to part (a).

Problem 2

A square coil with sides 8.0 cm is made from copper wire of radius 1.5 mm. A magnetic field perpendicular to the coil changes at a rate of $5.0 \, \text{mT s}^{-1}$. The resistivity of copper is $1.7 \times 10^{-8} \, \Omega \, \text{m}$.

Determine the current in the loop.

Problem 3

A vehicle with a radio antenna of length 0.85 m is travelling horizontally with a speed of $95 \, \text{km h}^{-1}$. The horizontal magnetic field due to the Earth where the vehicle is located is 0.055 mT.

a) Deduce the maximum possible emf that can be induced in the antenna due to the Earth's magnetic field.

b) Identify the relationship between the velocity of the vehicle and the magnetic field direction for this maximum to be attained.

Problem 4

Two identical circular coils, X and Y, are arranged parallel to each other and wound in the same sense. When direct current is switched on in X the current direction is clockwise in X.

Predict the direction of any induced current in Y when

a) the current in X is switched on.

b) the current in X is switched off.

Problem 5

a) A parallel plate capacitor is made from overlapping metal plates with an air gap in between.

State **two** ways of increasing the capacitance of the capacitor.

b) An RC circuit is constructed as shown.

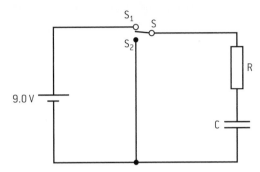

When the switch is moved from position $\textbf{S}_1$ to position $\textbf{S}_2$ the capacitor begins to discharge.

(i) The capacitance of the capacitor **C** is 470 µF. It takes a time of 60 s for the pd across the capacitor to fall to 5.0 V.

Calculate the resistance of the resistor **R**.

(ii) Sketch a graph to show the variation with time of the voltage across the capacitor, from the time when the switch goes to position $\textbf{S}_2$.

(iii) State and explain how you would modify the circuit so that it takes 90 s for the pd across the capacitor to reach 5.0 V rather than 60 s.

Problem 6

A 20 mF capacitor is connected to a 20 V supply. The capacitor can be discharged through a small motor. The energy of the capacitor can be used to lift a mass of 0.15 kg through a height of 0.80 m.

a) Calculate the initial energy stored by the capacitor.

b) Determine the efficiency of the energy conversion.

12 QUANTUM AND NUCLEAR PHYSICS (AHL)

12.1 THE INTERACTION OF MATTER WITH RADIATION

You must know:

✔ what is meant by the term photon

✔ what is meant by the photoelectric effect and that it cannot be explained by the wave theory of light

✔ that matter has a wave nature and that this can be demonstrated for electrons

✔ the Bohr model of the hydrogen atom and the quantization of angular momentum in the model

✔ that matter can have a wave function associated with it

✔ that there is a probability associated with the tunnelling of a particle through a potential barrier.

You should be able to:

✔ discuss the photoelectric experiment and solve problems involving the photoelectric effect

✔ explain which features of photoelectricity cannot be explained by classical wave theory

✔ explain what is meant by pair production and pair annihilation

✔ discuss experimental evidence for matter waves, including an experiment that demonstrates the wave nature of electrons

✔ state order-of-magnitude estimates from the uncertainty principle

✔ solve the uncertainty principle for energy–time and position–momentum.

When a zinc sheet is negatively charged and ultraviolet radiation is incident on it, the sheet loses negative charge. This does not occur with radiation of longer wavelengths or when the sheet is positively charged. The sheet is losing charge by electron emission.

Einstein's explanation of photoelectricity included the following points.

- Light of frequency f consists of *photons*, each with energy $E = h f$.

- h is the Planck constant.

- One photon incident on the metal can interact with one electron.

- No electrons are emitted when the photon energy is below a minimum value corresponding to a threshold frequency f_0.

- The minimum energy, known as the work function $\Phi \, (= h f_0)$, is associated with the energy required to overcome the attractive forces that oppose the removal of the electron from the metal.

- Any excess energy $(\Phi - h f_0)$ is transferred to the kinetic energy of the emitted electron.

- An increase in the incident light intensity increases the number of incident photons but does not change the energy of the photons. Therefore, an increase in intensity increases the number of electrons emitted per second—but does not change their maximum kinetic energy.

These led to the Einstein photoelectric equation $E_{max} = hf - \Phi$, which can also be written as $\frac{1}{2}m_e v_{max}^2 = hf - \Phi$, where v_{max} is the maximum speed of the electrons.

Figure 12.1.1 shows the experiment Millikan devised to test Einstein's result.

When monochromatic light with large enough photon energy is incident on the cell, the cathode emits electrons. Increasing the reverse pd between cathode and anode (anode negative) gives a pd at which electrons no longer reach the anode—the current falls to zero. The pd V_s at this point is known as the *stopping potential* and $eV_s = hf - \Phi$.

Figure 12.1.2 shows a graph of the variation of V_s with frequency $\frac{1}{\lambda}$.

The two different metals have the same gradient (which depends on h, c and e) but are offset depending on their work function.

Wave theory cannot account for photoelectricity as the effect is instantaneous. The wave theory suggests that, no matter how weak the light, given enough time, energy accumulates and releases an electron for all wavelengths. The photoelectric effect predicts that, when the photon energy is too small, no electron will ever be released and this is what is observed in practice.

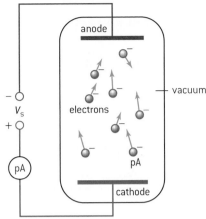

potential of anode is made negative so electrons cannot quite reach it and the picoammeter reading becomes zero

▲ **Figure 12.1.1.** Photocell for measuring h

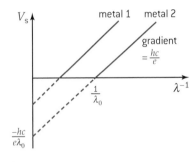

▲ **Figure 12.1.2.** The graph obtained in the Planck constant experiment for two different metals

Example 12.1.1

Electromagnetic radiation of frequency f is incident on a metal. The maximum kinetic energy $E_{k\,max}$ of photoelectrons from the surface is measured. The graph shows the variation with f of $E_{k\,max}$.

a) Calculate, using the graph, the Planck constant h.

b) Determine the minimum energy required to remove an electron from the metal surface.

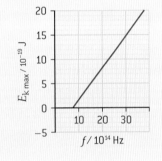

Solution

a) The equation for the graph is $E_{max} = hf - \Phi$.

Comparing this with $y = mx + c$ shows that the gradient (m) is h the Planck constant.

Reading off the coordinates $(30 \times 10^{14}, 14.5 \times 10^{-19})$ and $(7.5 \times 10^{14}, 0)$, the gradient is

$$\frac{(16 \times 10^{-19} - 0)}{(30 \times 10^{14} - 7.5 \times 10^{14})} = \frac{16 \times 10^{-19}}{22.5 \times 10^{14}} = 6.4 \times 10^{-34}\,\text{J s}$$

b) The minimum energy is the energy found by extrapolating the graph to the y-axis.

This value is $5.0 \times 10^{-19}\,\text{J}$.

(Alternatively, $\Phi = hf_0$ where f_0 is the intercept on the x-axis: $7.5 \times 10^{14}\,\text{J}$.)

>> **Assessment tip**

In Example 12.1.1 part a), the question begins 'Calculate, using the graph, the …'. In an exam, you *must* follow such instructions. The examiner will require the solution to be based on a graphical analysis. Few, if any, marks will be available for alternative approaches.

The **de Broglie wavelength** λ is given by $\lambda = \dfrac{h}{p}$, where p is the momentum of the particle.

Example 12.1.2

a) Calculate the Broglie wavelength of an electron mass 9.1×10^{-31} kg and speed 9.0×10^6 m s^{-1}.

b) Explain why electrons of this speed can be used to investigate crystal structures using electron diffraction.

Solution

a) Use $\lambda = \dfrac{h}{m_e v}$

$$= \frac{6.6 \times 10^{-34}}{9.1 \times 10^{-31} \times 9.0 \times 10^6}$$

$$= 8.1 \times 10^{-11} \text{ m}$$

b) The separation of atoms is of the order of 10^{-10} m, so is roughly the same as the wavelength of the electron found in part a).

🔗 The basics of the Rutherford–Geiger–Marsden experiment are covered in Topic 7.3.

The Bohr theory leads to the suggestion that $mvr = \dfrac{nh}{2\pi}$ and that the energy, in eV, of a particular orbital is given by $E = -\dfrac{13.6}{n^2}$.

🔗 This prediction fits extremely well with the energy level changes outlined in Topic 7.1. However, the theory was a poor fit for atoms more complex than hydrogen and helium.

Light has some wave-like properties and some particle-like properties. This leads to the idea of *wave-particle duality*. In 1924, French physicist Louis de Broglie suggested that this duality applied also to matter. In other words, under certain circumstances, matter could be observed to demonstrate wave-like properties as *matter waves*.

He suggested that particles have a *de Broglie wavelength*.

This hypothesis was verified by the American physicists, Clinton Davisson and Lester Germer. They diffracted a beam of electrons by a crystal of nickel. The electrons, after diffraction, form a series of bright and dark fringes on the front of the tube.

The wavelength of an electron that has been accelerated from rest through a potential difference V, travelling at non-relativistic speed, can be calculated by equating the kinetic energy to the electrical energy: $\dfrac{1}{2}m_e v^2 = eV$ and $m_e v = p = \sqrt{2m_e eV}$. So $\lambda = \dfrac{h}{\sqrt{2m_e eV}}$.

In 1913, Danish physicist Niels Bohr proposed a model of the atom to account for the observations of Geiger and Marsden and to address the problem that the concept of an electron orbiting a nucleus contradicted classical physics. Bohr proposed that:

- electrons exist in certain stationary states without emitting any radiation (an orbiting electron does this according to classical physics)

- electrons emit or absorb radiation of frequency f only when moving between stationary states (the energy E transferred is equal to hf)

- the angular momentum of an electron in a stationary state is quantized with values $\dfrac{nh}{2\pi}$, where n is an integer.

The third assumption is equivalent to fitting an integral number of wavelengths as a standing wave onto the electron orbit.

Austrian physicist Erwin Schrödinger made the next breakthrough in understanding. He described the quantum states of particles using a wave function ψ that gives an amplitude for each position. ψ^2 is proportional to the probability density P(r). Mathematically, $P(r) = |\psi|^2 \Delta V$, where $P(r)$ is the probability of finding a particle a distance r from an origin and ΔV is the size of the volume considered.

One way to align the wave and particle approaches is to consider the electron standing waves that exist in the negative potential well for the electron–proton system of the hydrogen atom. Figure 12.1.3 shows the probability functions for four states of the atom in a simple one-dimensional version.

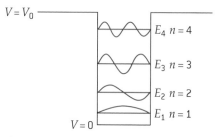

▲ Figure 12.1.3. The probability functions for four energy states of a simple one-dimensional atom

All the wave functions have a node at the atom edge suggesting that the probability of finding the electron there is zero. The antinode positions of the standing waves give the maximum probability density.

Probability functions introduce the concept of Heisenberg's uncertainty in measurement and observation. First introduced in 1927 by German physicist Werner Heisenberg, it places a limit on the precision with which a measurement of two quantities (known as conjugate variables) can be made. Heisenberg suggested that the uncertainty of position Δx and uncertainty of momentum Δp are given by $\Delta x \Delta p \geq \dfrac{h}{4\pi}$. There is a corresponding relationship between energy E and time t: $\Delta E \Delta t \geq \dfrac{h}{4\pi}$.

Where electric fields are strong (near an atomic nucleus, for example) a photon with sufficient energy can spontaneously convert into a particle–antiparticle pair. This is known as *pair production* and typically leads to the creation of an electron together with a positron or a proton–antiproton pair. The incoming photon must have enough energy to create both particles.

Energy requirements mean that photon energy is $E = 2mc^2$, where m is the mass of one of the particles being created. The rest energy of an electron is 0.511 MeV. Therefore, the minimum energy of the photon must be 1.02 MeV.

The positron from the pair production is likely to be short lived. It will soon meet another electron and annihilate, forming two photons. The total energy of both emitted photons must equal the total mass–energy of the annihilating particles.

Under some circumstances, production of an electron–positron pair is possible with much less than the 1.02 MeV of energy—but only for a short time. The uncertainty principle $\Delta E \Delta t \geq \dfrac{h}{4\pi}$ predicts that a low-energy photon near to an atomic nucleus can create two particles with short lifetimes.

The wave function of a particle suggests that the particle has a finite probability of being at any place in the Universe at one time. This probability may be very small but it is non-zero. This accounts for the phenomenon of alpha-particle decay.

The alpha particle (2 protons + 2 neutrons) is a particularly stable grouping of nucleons (see, for example, the position of helium on the binding energy per nucleon chart in Topic 7.2). There is a tendency for the four particles to group within the nucleus.

Figure 12.1.4 shows what can happen next. The energy of the alpha particle within the nucleus is not enough to escape over the Coulomb energy barrier because there is an average attractive force on all nucleons within the nucleus. Alpha particles emitted in the decay tunnel through the energy barrier, with a reducing amplitude inside the barrier and a much reduced—though finite—amplitude of the wave function outside.

The existence of this wave function beyond the barrier means that alpha particles escape. Quantities such as the radioactive decay constant for the alpha decay of an individual nuclide are determined by the relative barrier heights and alpha energies.

> The pair production must conserve charge, lepton number, baryon number and strangeness. These conservation laws were considered in Topic 7.3.

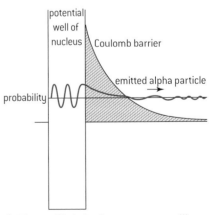

▲ Figure 12.1.4. Quantum tunnelling of alpha particles

Example 12.1.3

An electron is confined within a region of length 2.0×10^{-10} m.

The variation with distance x of the wave function Ψ of the electron at a particular time is given.

a) Outline what is meant by *wave function*.

b) Estimate the momentum of the electron.

c) Deduce the uncertainty in the momentum of the electron.

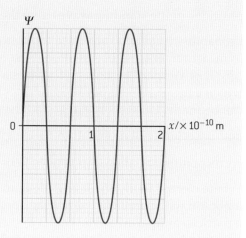

Solution

a) The wave function is the property of a particle. It is proportional to the square root of the probability of finding the particle at a specified position.

b) There are three wavelengths of the electron wave function in 2.0×10^{-10} m.

Therefore wavelength $= \dfrac{2.0 \times 10^{-10}}{3} = 6.7 \times 10^{-11}$ m.

So $p = \dfrac{h}{\lambda} = \dfrac{6.6 \times 10^{-34}}{6.7 \times 10^{-11}} = 1.0 \times 10^{-23}$ N s

c) The uncertainty in the position of the electron is 2.0×10^{-10} m.

Heisenberg uncertainty gives $\dfrac{6.6 \times 10^{-34}}{4\pi \times 2.0 \times 10^{-10}} = 2.6 \times 10^{-25}$ N s

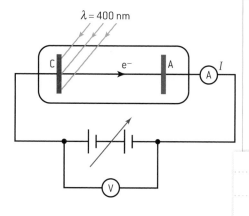

λ = 400 nm

An apparatus is used to investigate the photoelectric effect. A caesium cathode C is illuminated by a variable light source. A variable power supply is connected between C and the collecting anode A.

A current is observed on the ammeter when violet light illuminates C. With V held constant the current becomes zero when the violet light is replaced by red light of the same intensity. Explain this observation. [3]

This answer could have achieved 2/3 marks:

According to the quantum model of light where light is made up of photons (E = h f), the energy provided by the light is dependent on the frequency of light. When red light, which has lower frequency, shines on the plate, the energy is not enough to remove the electrons. Therefore, electrons don't flow from C to A → no current → 0A. Violet light has higher frequency, which is enough to liberate the electrons off the plate C to travel to plate A → There's a current.

▼ This question is all about comparison. The answer needs to be clearer about the *differences* between red and violet light and what these differences lead to. It can be inferred that the violet energy is greater than the red energy and this mark was awarded—but it is not the job of the examiner to do work for a candidate. There is no link to the work function—the term is not mentioned. There is no link between the energy of a red light photon and the fact that this is less than the work function.

▲ There is a statement that indicates that the light arrives as photons and that each photon has an energy $E = hf$.

This answer could have achieved 0/3 marks:

When violet light, which has lower wavelength than red light, is used, the electrons have kinetic energy given by KE = hf - φ, where φ is work function of caesium. Since violet light has a lower wavelength than red light, the electrons emitted when violet light is shone on the cathode have greater kinetic energy than the electron emitted when red light is shone on the cathode. Hence, while the electrons have sufficient kinetic energy to overcome electrostatic forces of repulsion from the anode when violet light is shone on the cathode. Electrons that are emitted when red light is shone on the cathode do not have sufficient kinetic energy to do so. This is why electrons cannot reach the anode when violet light is replaced with red light, and subsequently why current is zero when violet light is replaced by red light.

▼ This answer is entirely in terms of the kinetic energy of the electrons. An attempt is made to explain Einstein's photoelectric equation, but again this is wrong as it is in terms of electrons.

12.2 NUCLEAR PHYSICS

You must know:

✔ that Rutherford scattering leads to an estimate of nuclear radius

✔ that a nucleus has nuclear energy levels

✔ the properties of the neutrino and how its existence was deduced using observations from beta decay

✔ the law of radioactive decay

✔ that the probability of decay of a nucleus per unit time is the decay constant.

You should be able to:

✔ estimate the radius of a nucleus using the rule that $R \propto A^{\frac{1}{3}}$

✔ describe a scattering experiment that uses the minimum intensity location based on the de Broglie wavelength

✔ explain deviations from Rutherford scattering at high energies

✔ solve problems involving radioactive decay for time intervals that are not integer half-lives

✔ explain methods for measuring short and long half-lives.

Head-on scattering means that the alpha particle returns along the same path by which it arrived. The point at which the alpha particle changes direction is where all its initial kinetic energy is transferred to electric potential energy stored in the nucleus–alpha system.

$\frac{1}{2}m_\alpha v_\alpha^2 = \frac{k(2e)Ze}{r^2}$, where r is the distance of closest approach, m_α is the mass of the alpha particle, v_α is the initial speed of the alpha particle, Z is the proton number of the nucleus and k is the Coulomb constant.

This equation assumes that the nucleus does not recoil. The larger v_α, the closer the alpha particle will be to the nuclear centre before deflection.

🔗 Topic 7.3 discussed Rutherford scattering and showed that the following deductions can be made from the observations of the Geiger–Marsden experiment.

• Most of the atom is empty space.

• There are small dense regions of positive charge in the atom.

The Coulomb constant is k in the formula $F = \dfrac{kq_1q_2}{r^2}$; see Topic 5.1.

The **radius R of a nucleus** with nucleon number A: $R = R_0A^{\frac{1}{3}}$, where R_0 is a constant with value 1.20 fm (1.20×10^{-15} m).

Example 12.2.2

The radius of the $^{12}_{6}$C nucleus is 3.0×10^{-15} m. Calculate the radius of the $^{20}_{10}$Ne nucleus.

Solution

$R = R_0A^{\frac{1}{3}} \Rightarrow$

$R_{Ne} = R_C \times \left(\dfrac{A_{Ne}}{A_C}\right)^{\frac{1}{3}}$

$= 3.0 \times 10^{-15}\left(\dfrac{20}{12}\right)^{\frac{1}{3}}$

$= 3.6 \times 10^{-15}$ m

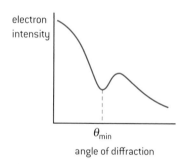

▲ **Figure 12.2.1.** Intensity–diffraction angle graph for electron scattering

Example 12.2.1

a) Determine the kinetic energy of an alpha particle travelling at 2.00×10^7 m s^{-1}. Ignore relativistic effects.

b) Calculate the closest distance of approach for a head-on collision between the alpha particle and a gold nucleus, $^{197}_{79}$Au. Assume that the gold nucleus does not recoil.

Solution

a) The mass of the alpha particle $= 4.0026 \times 1.66 \times 10^{-27}$
$= 6.64 \times 10^{-27}$ kg

Its kinetic energy $= \dfrac{1}{2}mv^2 = \dfrac{6.64 \times 10^{-27} \times \left(2.00 \times 10^7\right)^2}{2} = 1.33 \times 10^{-12}$ J

b) The gain in electric potential energy when the nucleus and alpha particle are at their closest separation r must equal the loss in kinetic energy. The alpha particle charge is $+2e$ and the gold nucleus charge is $+Ze$.

Gain in potential energy $= 1.33 \times 10^{-12} = \dfrac{kq_1q_2}{r} = \dfrac{k \times (2e) \times Ze}{r}$

Substituting gives $\dfrac{k \times 2 \times 79 \times \left(1.6 \times 10^{-19}\right)^2}{r}$ so $r = 2.7 \times 10^{-14}$ m

Nuclear radius measurements show that nuclear density is roughly constant whatever the nucleon number A of the nuclide. This density is very high, approximately 2×10^{17} kg m^{-3}. This value is approached only by that of a neutron star—a matchbox-full of which has a mass of three-billion tonnes.

The expression for the radius of a nucleus is derived from the fact that nuclear density ρ is constant. Since $\rho = \dfrac{AM}{\frac{4}{3}\pi R^3}$, where M is the mass of a nucleon, $R^3 = \dfrac{3M}{4\pi\rho}A$ and $R = \sqrt[3]{\dfrac{3M}{4\pi\rho}}A^{\frac{1}{3}}$.

Beams of protons or electrons can be used to probe the nucleus using diffraction. Figure 12.2.1 shows a typical variation of scattered electron intensity with scattered (diffracted) angle for a metal. There is a pronounced minimum at θ_{min}.

This angle is related to the diameter D of the atom by $\sin\theta = \dfrac{\lambda}{D}$, where λ is the de Broglie wavelength of the scattered particle.

Deviations from Rutherford scattering are seen when high-energy electrons (> 420 MeV or so) are scattered in experiments. These deviations occur because:

- the collisions become inelastic; the incident electrons lose kinetic energy that is transferred into mass as mesons are created and emitted from the nucleus

- deep inelastic scattering occurs as the electrons penetrate further into the nucleus; they then scatter off quarks inside nucleons (this type of scattering has provided evidence for the Standard Model).

Example 12.2.3

a) Explain, with quantitative detail, why electrons of energy 400 MeV are used in determining nuclear size. Ignore relativistic effects in your estimate.

b) Electron diffraction experiments lead to information about nuclear density and the average separation of nucleons.

Outline the main conclusions drawn for:

i) nuclear density

ii) average separation of particles.

Solution

a) $\lambda = \dfrac{hc}{E} = \dfrac{6.6 \times 10^{-34} \times 3.0 \times 10^{8}}{400 \times 10^{6} \times 1.6 \times 10^{-19}} \approx 3$ fm

The de Broglie wavelength for the electrons of this energy is about 10^{-15} m which is of the same order of magnitude as a nuclear diameter. This means that the 400 MeV electrons have a diffracting wavelength comparable to the size of their target.

b) i) Nuclear density is constant; the relationship $R \propto A^{\frac{1}{3}}$ can be written as $R^3 \propto A$.

This leads to $\dfrac{\text{volume}}{\text{mass}} = \text{constant}$.

ii) As a consequence of the constant density, the average separation of the nucleons must also be constant.

Nuclei, like atoms with their energy states, have nuclear energy levels that are observed when daughter nuclei are emitted in radioactive decay. The gamma radiation is emitted at a small number of discrete energies.

Some of the daughter nuclei are formed in a metastable state. The nucleus remains in this excited state for a short time before emitting the remainder of its energy and moving to the ground state. This release is involved in the decay of technetium, a nuclide used in medical diagnosis.

The full decay from unstable molybdenum (Mo) that involves technetium (Tc) is

$^{99}_{42}\text{Mo} \rightarrow {}^{99m}_{43}\text{Tc}^{*} + {}^{0}_{-1}\text{e} + \bar{\upsilon}_{\text{e}}$ and $^{99m}_{43}\text{Tc}^{*} \rightarrow {}^{99}_{43}\text{Tc} + \gamma$ (99m and * mean a metastable state).

Alpha decay and beta (β^{-}) decay are very different in energy terms. The alpha particles emitted show only one or two energy values.

In alpha emission, one active nucleus decays into the daughter nucleus and the alpha particle. When the initial momentum before decay is zero, the final momentum must also be zero. The alpha particle and the daughter nucleus move in opposite directions with speeds in the ratio

$\dfrac{\upsilon_{\alpha}}{\upsilon_{\text{d}}} = \dfrac{m_{\text{d}}}{m_{\alpha}}$.

In beta-minus decay, emitted electrons are observed to have a complete range of energies from zero to a maximum that is slightly less than the maximum energy believed to be available (Figure 12.2.2).

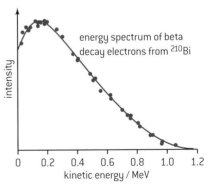

energy spectrum of beta decay electrons from ^{210}Bi

intensity

kinetic energy / MeV

▲ Figure 12.2.2. The energy spectrum from the decay of Bismuth-210

Topic 7.1 describes radioactive decay as a *random* and *spontaneous* process in which an individual nucleus decays into a daughter nucleus with the emission of particles.

Italian physicist Enrico Fermi interpreted this to mean that three objects are produced in the decay: the daughter nucleus, the beta particle and an unknown particle. The three particles can share the available energy in many ways, leading to a beta energy spectrum.

In 1933, Fermi named this particle the neutrino, meaning "**little neutral one**". Experiments confirmed this prediction and the neutrino has since been indirectly observed.

Neutrinos are difficult to observe. Large numbers emitted in fusion reactions in the Sun pass through the Earth every second. The particles can only be observed when a few interact indirectly with nuclei under special circumstances.

The properties of the neutrino include neutral charge and effectively zero mass.

The rate of change of nuclei with time $\left(\dfrac{dN}{dt}\right)$ is the activity A of the sample.

This equation can be written as

$$A = \frac{dN}{dt} = -\lambda N \text{ or } \frac{dN}{N} = -\lambda dt,$$

where A is the activity (decays per second) and this second equation leads to the solution $N = N_0 e^{-\lambda t}$, where t is the time and N_0 is the initial number of undecayed nuclei at $t = 0$.

The activity A of the sample is $A = \lambda N_0 e^{-\lambda t}$.

Example 12.2.4

a) Calculate, in joules, the maximum beta-minus particle energy from Figure 12.2.2.

b) All the beta particles emitted by Bismuth-210 arise from identical energy changes in the bismuth nucleus.

Explain how Figure 12.2.2 suggests that an anti-electron neutrino must also be emitted.

Solution

a) The maximum energy is 1.2 MeV.

This corresponds to $1.2 \times 10^6 \times 1.6 \times 10^{-19}$ J $\equiv 1.9 \times 10^{-13}$ J.

b) The total energy available from the decay is constant. The existence of a beta energy spectrum must mean that there is no unique way to distribute the energy between the beta particle and the daughter product.

There are an infinite number of ways to distribute the energy three ways. This is an indication that three particles are involved. The energy is shared between the daughter nucleus, the electron and the antineutrino.

>> **Assessment tip**

The negative sign in $\dfrac{dN}{dt} = -\lambda N$ arises because the number of undecayed nuclei decreases with time.

Each identical nucleus of a particular nuclide has the same probability of decay per unit time. This probability is known as the *decay constant* λ; it provides a fundamental relationship of radioactive decay:

rate of loss of nuclei $= -\lambda \times$ number of nuclei remaining

The relationship between λ and half-life $t_{\frac{1}{2}}$ is an important one. After one half-life, the number of atoms has halved. So $\dfrac{N_0}{2} = N_0 \exp(-\lambda t_{\frac{1}{2}})$.

Eliminating N_0 and taking logarithms gives $\ln 0.5 = -\lambda t_{\frac{1}{2}}$.

This can be written as $t_{\frac{1}{2}} = \dfrac{\ln 2}{\lambda}$.

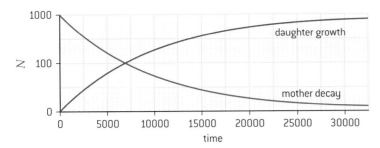

▲ Figure 12.2.3. Decay and growth curves for stable daughter

Figure 12.2.3 shows how N varies with t both for the nuclei that are decaying and the daughter nuclei that are forming. The graph assumes that the daughter nuclide is stable and does not undergo further radioactive decay.

The determination of λ for nuclides with long half-lives involves obtaining the nuclide in a pure sample and then measuring the mass of the sample. This leads to the number of atoms N in the sample. Then the total activity A of the sample can be measured (this involves making an estimate using the counts collected by a detector of finite size and factoring this up to the counts over a complete sphere surrounding the sample).

The equation $A = \lambda N$ is used to calculate the decay constant.

Reasonably short half-lives can be measured by:

• measuring the background count rate in the laboratory

• taking readings of count rate against time until the value equals that of the background

• subtracting the background from each reading

• assuming that the corrected count rate $\propto$ activity

• plotting a graph of $\ln A$ against time

• finding the gradient of this graph, which gives $-\lambda$.

Example 12.2.5

Radioactive iodine (I-131) has a half-life of 8.04 days.

a) Calculate, in seconds^{-1}, the decay constant of I-131.

b) Calculate the number of atoms of I-131 required to produce a sample with an activity of 60 kBq.

c) Deduce the time taken for the activity of the sample in part b) to decrease to 15 kBq.

Solution

a) $\lambda = \dfrac{\ln 2}{8.04 \times 24 \times 60 \times 60} = 1.0 \times 10^{-6}\ \text{s}^{-1}$

b) The number of atoms is given by $N = \dfrac{A}{\lambda} = \dfrac{6.0 \times 10^{4}}{1.0 \times 10^{-6}} = 6.0 \times 10^{10}$

c) A decrease from 60 to 15 kBq is $\dfrac{1}{4}$; in other words, two half-lives, which is 16 days.

Rhodium-106 $\left(^{106}_{45}\text{Rh}\right)$ decays into palladium-106 $\left(^{106}_{46}\text{Pd}\right)$ by beta minus (β^-) decay. The diagram shows some of the nuclear energy levels of rhodium-106 and palladium-106. The arrow represents the β^- decay.

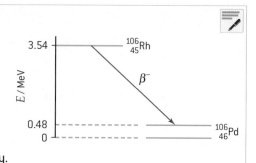

a) Explain what may be deduced about the energy of the electron in the β^- decay. [3]

This answer could have achieved 2/3 marks:

The energy of the β^- particle is 3.54 – 0.48 MeV = 3.06 MeV. The energy is smaller than 3.54 MeV because it shares its energy with an antineutrino. Thus energy < 3.06 MeV.

▲ The answer shows a correct subtraction that leads to the maximum energy for the beta particle (3.54 – 0.48 = 3.06 MeV). There is also the recognition that this is a maximum energy for the beta particles.

▼ The answer fails to make it clear that the minimum beta energy is 0.

b) Suggest why the β^- decay is followed by the emission of a gamma ray photon. [1]

This answer could have achieved 1/1 marks:

The palladium nucleus is not in its ground state and it releases the gamma ray to go to its ground state.

▲ The answer correctly recognizes that the nucleus is left in an excited state and loses the excess energy via the emission of a gamma photon.

c) Calculate the wavelength of the gamma ray photon in part b). [1]

This answer could have achieved 1/1 marks:

$$\lambda = \frac{hc}{E} = \frac{1.24 \times 10^{-6}\ eVm}{0.48\ MeV} = 2.583 \times 10^{-12}\ m \approx 2.6 \times 10^{-17}\ m$$

▲ The calculation is clear and accurate.

Practice problems for Topic 12

Problem 1

Electromagnetic radiation incident on a metal causes a photoelectron to be emitted from the surface.

a) State and explain **one** aspect of the photoelectric effect that suggests the existence of photons.

b) The work function of sodium is 2.3 eV.

(i) Outline what is meant by work function.

(ii) Electromagnetic radiation of wavelength 320 nm is incident on sodium.

Determine the maximum kinetic energy of the electrons emitted from the sodium.

Problem 2

Electrons are emitted from a heated cathode and accelerated in a vacuum through a potential difference as a narrow beam. This beam is fired at a polycrystalline graphite target in a chamber. The inside surface of the chamber is coated with fluorescent material that emits light when the electrons release their energy to it.

a) The electrons reach the inside surface travelling at a speed of $4.0 \times 10^7 \, \text{m s}^{-1}$.

Calculate the de Broglie wavelength of the electrons.

(i) Sketch the pattern of light you would expect to see emitted by the fluorescent material.

(ii) Explain why the pattern suggests that electrons have wave-like properties.

b) Explain **one** aspect of the experiment that suggests that electrons have particle-like properties.

Problem 3

Radioactive sodium $\left({}^{22}_{11}\text{Na} \right)$ has a half-life of 2.6 years. A sample of this nuclide has an initial activity of $5.5 \times 10^5 \, \text{Bq}$.

a) Explain what is meant by the *random nature* of radioactive decay.

b) Sketch a graph of the activity of the sodium sample for a time period of 6 years.

c) Calculate:

(i) the decay constant, in s^{-1}, of ${}^{22}_{11}\text{Na}$

(ii) the number of atoms of ${}^{22}_{11}\text{Na}$ in the sample initially

(iii) the time taken, in s, for the activity of the sample to fall from 100 kBq to 75 kBq.

Problem 4

Electrons are emitted instantaneously from a metal surface when monochromatic light of wavelength 420 nm is incident on the surface.

a) Explain why the energy of the emitted electrons does not depend on the intensity of the incident light.

b) Suggest why the electron emission is instantaneous.

c) The work function of the metal is 2.3 eV and one electron is emitted for every 4500 photons incident on the surface. The surface area of the metal is $4.5 \times 10^{-6} \, \text{m}^2$.

(i) Determine, in J, the maximum kinetic energy of an emitted electron.

(ii) Determine the initial electric current from the surface when the intensity of the incident light is $3.8 \times 10^{-6} \, \text{W m}^{-2}$.

Problem 5

Caesium-137 $\left({}^{137}_{55}\text{Cs} \right)$ decays by negative beta decay to form a nuclide of barium (Ba).

a) Write down the nuclear reaction for this decay.

b) The half-life of caesium-137 is 30 years. Determine the fraction of the original caesium that remains after 200 years.

c) Caesium is a waste product in a nuclear reactor.

Suggest why the fuel rods in the reactor are removed well before the uranium is completely converted.

Problem 6

The radius of a carbon-12 $\left({}^{12}_{6}\text{C} \right)$ nucleus is $3.1 \times 10^{-15} \, \text{m}$.

a) Determine the radius of a magnesium-24 $\left({}^{24}_{12}\text{Mg} \right)$ nucleus.

b) Sketch a graph to show the variation of nuclear radius with nucleon number.

Annotate your graph with both the C-12 and Mg-24 nuclei.

13 DATA-BASED AND PRACTICAL QUESTIONS (SECTION A)

You must know:

✔ what is meant by random and systematic errors in measurements.

You should be able to:

✔ plot points accurately on a graph

✔ construct error bars on a graph

✔ sketch best-fit lines on graphs

✔ determine and interpret the meaning of gradients and intercepts on graphs

✔ determine the errors in gradients and intercepts.

Paper 3, Section A contains data-analysis questions. The skills you need to answer these questions are closely linked to the practical skills you demonstrate in the *internal assessment* and those described in Topic 1. This chapter provides guidelines that you will find helpful in all your practical work for the DP physics course.

Communication using a graph is essential in physics; graphs allow relationships between data points to be grasped visually and quickly. It helps if you know the best graph to plot to display your data. First, consider how best to render your data as a straight line (in the form $y = mx + c$, where y and x are the data, m is the gradient of the line and c is the intercept on the y-axis). This may involve algebraic manipulation.

> There is a table in the Internal Assessment chapter that lists many of the relationships you meet in the course and how you should manipulate the data to obtain a straight-line trend.

Example 13.1

A simple pendulum is suspended from the ceiling of height D above the floor. There is no access to the suspension at the ceiling. The period T of oscillation of the pendulum is measured as a function of the vertical distance h from the floor to the pendulum bob.

Suggest a suitable graph to display the data.

Solution

The length of the pendulum is $D - h$.

The equation for the period is $T = 2\pi\sqrt{\dfrac{l}{g}} = 2\pi\sqrt{\dfrac{D-h}{g}}$.

A plot of T against h will not be a straight line.

However, squaring both sides and rearranging gives

$$T^2 = -\frac{4\pi^2}{g}h + \frac{4\pi^2}{g}D.$$

This is now in the form $y = mx + c$.

So a plot of T^2 against h will give a straight line of gradient $-\dfrac{4\pi}{g}$ with an intercept on the y-axis of $\dfrac{4\pi}{g}D.$

Once you have collected data and identified a suitable graph, the graph can be plotted.

Here are some guidelines for drawing graphs.

- Plotted graphs need **labels** and **units** on both axes; for example, speed / ms^{-1}. The / does not mean *divide*, it means *measured in*. You should write units in the form ms^{-1} rather than m/s.

- **Scales** on the axes should be straightforward using 1:2, 1:5 or 1:10. The ratios 1:3, 1:6, 1:7 or 1:9 should never be used. Only use 1:4 when absolutely necessary.

- Always **fill** as much of the printed grid as possible. Aim to have a minimum range for your plots of half the grid area. This may mean using a false origin (one that does not begin at $(0,0)$).

- **Data points** should be marked clearly using × or +.

- All markings on a graph should be in **sharp, black pencil** (so that you can erase mistakes). As a guideline, when your pencil line is as thick as the thickest lines on the grid, the pencil is not sharp enough.

- Draw all straight lines with a **ruler** (preferably transparent).

- Use **freehand for curves**. Practise the curve several times first, drawing with your hand inside the curve, and then draw the line in one continuous movement.

- Try to get the **same number of points** on both sides of the line, minimizing the total distance of all points from the line.

- Do not force your line to go through the origin unless you have a very good reason to distort your line in this way.

Modern calculators have internal programs that can determine the slope and intercept of a line given the data points. If your calculator can do this, then use this facility. Another quick way to find the best straight line is to divide your data into two groups: upper and lower (Figure 13.1). Find the separate x and y averages of both groups and plot these two points. Join them and that is your straight line. The gradient is easily obtained from the two mean values too.

Once your graph is plotted, many more data are available to you. Remember that, besides being a way of visualizing your data, a graph is an averaging technique from which you can obtain:

- the **gradient of a straight line**, by calculating change in y-coordinates ÷ change in x-coordinates

- the **gradient at a point on a curve**, by drawing the tangent at the point concerned and calculating the gradient of this new straight line. To draw this tangent, use a small plane mirror and align it so that when looking into it the curve and its reflection appear continuous. The mirror is then at 90° to the curve, draw along the mirror—a protractor gives the tangent line directly from this.

- The **area under a graph** can be found either algebraically or can be made into a series of triangles and rectangles when the line is straight. When it is a curve, the area can be found by counting squares and determining the area value of each square.

>> **Assessment tip**

A common question is 'Draw the best-fit line'. This does *not* mean draw the best-fit straight line, as the best fit line could be curved with the data points distributed evenly about it.

>> **Assessment tip**

Draw the lines from which you will calculate a gradient as long as possible so they stretch from axis to axis. Then use at least half of this line for the determination. The longer the line, the smaller the fractional uncertainty in your gradient result.

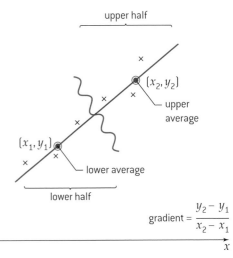

▲ **Figure 13.1.** Dividing plots into two groups to find an approximate best-fit line

An **error bar** is an I-shape or H-shape centred on the datum point. The width or height of the bar indicates the error in the measurement. The nature of the experiment will determine whether each point on the graph has the same or different absolute errors.

Topic 1.2 describes the treatment of errors in single data points whether absolute, fractional or percentage.

In Paper 3, Section A, error bars may be used and you will normally be told when the errors in one measurement are negligible.

If you are studying at Higher Level you will need to be able to plot and analyse log–log and log–linear graphs (eg in radioactive decay the variation of ln(activity / Bq) with time). Notice how the unit for activity is inside the bracket associating it with the *quantity*, not with its *logarithm*.

- The **intercept** on either axis can be found either by directly reading it off or by using trigonometry to calculate the intercept when there is a false origin. This will almost certainly involve extrapolation of your plotted line to the intercept.

Every data point on your graph will have an error associated with it. When the absolute error in the point is much less than the size of the grid you need do nothing further. However, when the absolute error is roughly the size of the smallest square or larger, then you need to consider the use of *error bars*.

When every point on a graph has an error bar, it is possible to construct the maximum and minimum gradients. The construction is shown in Figure 13.2. The two lines must lie on the outer edges of the relevant error bars. Once they have been drawn, the two gradients can be calculated in the usual way. The absolute error in the gradient is

$$\frac{\text{maximum gradient} - \text{minimum gradient}}{2}.$$

This is similar to the calculation of an absolute error from the overall range of data for a particular datum point.

The absolute error in the gradient can then be used in further error calculations.

Similar ideas will apply to the intercepts on the axes. In Figure 13.2, the intercept on the y-axis is $(108 \pm 5)\,\text{m s}^{-1}$.

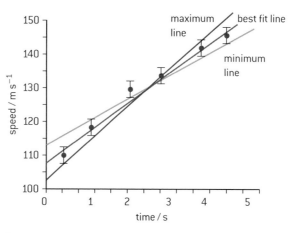

▲ **Figure 13.2.** Minimum and maximum gradients on a graph

A student suggests that the relationship between I and x is given by:

$$\frac{1}{\sqrt{I}} = Kx + KC$$

where K and C are constants.

Data for I and x are used to plot a graph of the variation of $\frac{1}{\sqrt{I}}$ with x.

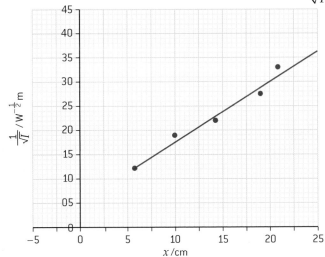

a) Estimate C. [2]

This answer could have achieved 0/2 marks:

y intercept is (0, 5)

C = 5m

▼ The y-intercept is correct. However, the candidate was asked for the value of C. When $\frac{1}{\sqrt{I}} = 0$, $x = C$ and this is the intercept on the x-axis.

b) Determine P to the correct number of significant figures including its unit, where $P = \frac{4\pi}{K^2}$. [4]

This answer could have achieved 4/4 marks:

From (b)(i), the gradient $K = \dfrac{30-5}{20-0} = 1.3 \, W^{-\frac{1}{2}}$ by considering (0, 5), (20, 30)

From (a), $K = 2\sqrt{\dfrac{\pi}{P}}$ so $\dfrac{K}{2} = \sqrt{\dfrac{\pi}{P}}$

$$\frac{K^2}{4} = \frac{\pi}{P} \Rightarrow P = \frac{4\pi}{K^2}$$

Therefore, $P = \dfrac{4\pi}{K^2} = 8.0 \, W \, m^{-2} \, cm^2$

$= 8.0 \times 10^{-4} \, W \, m^{-2} m^2$

$= 8.0 \times 10^{-4} \, W$

▲ There are a number of good points about this answer: it is clear and well presented. The candidate makes the source of the data for the gradient clear and the data points concerned are far apart and on the line. The calculations are correct and the unit and number of significant figures are correct too.

c) Explain the disadvantages that a graph of I versus $\frac{1}{x^2}$ has for the analysis in part (a) and part (b). [2]

This answer could have achieved 1/2 marks:

Since slope will be $= \dfrac{1}{K^2}$, it will be too small to measure from graph to right no. of significant figures.

Y intercept will be too small → hard to measure c

▼ There is some credit given for the point about the gradient. However, this is not the whole story: the equation becomes $I = \dfrac{1}{K^2(x+C)^2}$ so, unless C is known, the graph cannot be plotted to give a straight line.

A student measures the refractive index of a glass microscope slide.

He uses a travelling microscope to determine the position of x_1 of a mark on a sheet of paper. He then places the slide over the mark and finds the position x_2 of the image on the mark when viewed through the slide. Finally he uses the microscope to determine the position of x_3 of the top of the slide.

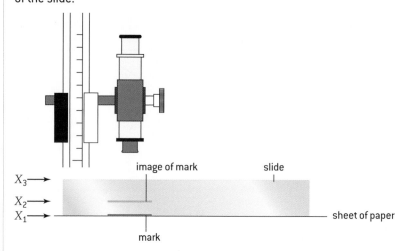

The table shows the average results of a large number of repeated measurements.

	Average position of mark / mm
x_1	0.20 ± 0.02
x_2	0.59 ± 0.02
x_3	1.35 ± 0.02

a) The refractive index of the glass from which the slide is made is given by:

$$\frac{x_3 - x_1}{x_3 - x_2}$$

Determine:

i) the refractive index of the glass to the correct number of significant figures, ignoring any uncertainty. [1]

This answer could have achieved 0/1 marks:

▲ After an initial error, the remainder of the working is carried through correctly.

$$\frac{1.35 - 0.2}{1.35 - 0.59} = \frac{1.15}{0.76} = 1.51$$

ii) Determine the uncertainty of the value calculated in part **i**. [3]

This answer could have achieved 2/3 marks:

▼ There are subtractions to find the differences in the image position but the uncertainties need to be obtained correctly. As this is a subtraction, the absolute uncertainties need to be added. The absolute errors in $x_3 - x_1$ and $x_3 - x_2$ are $0.02 + 0.02$ not 0.02 alone as the solution indicates.

$$\frac{\Delta h}{n} = \frac{\Delta x}{x_3 - x_1} + \frac{\Delta x}{x_3 - x_2} = \frac{0.02}{1.15} + \frac{0.02}{0.76}$$

$\Delta n = 1.5 \, (0.0437)$ where n is the refractive index

$\Delta n \approx \pm 0.0655 = \pm 0.1 = \Delta n$

b) After the experiment, the student finds that the travelling microscope is badly adjusted so that the n measurement if each position is too large by 0.05 mm.

Outline the effect that this error will have on the calculated value of the refractive index of the glass. [2]

This answer could have achieved 2/2 marks:

It will have no effect because this error will be cancelled out once $n = \dfrac{x_3 - x_1}{x_3 - x_2}$ is calculated.

▲ The point is that the refractive index is calculated by *differences*. Each measurement in $\dfrac{x_3 - x_1}{x_3 - x_2}$ is changed by the same amount and the overall ratio does not change. This could have been expressed more clearly in the answer.

Practice problems for Section A

Problem 1

Data are obtained for the variation in the extension of a thin filament when a force is applied to the filament. A graph of the results is shown.

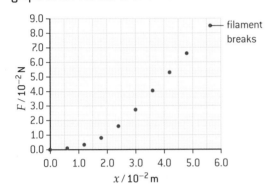

a) Draw a best-fit line for this graph.

b) The stress in the filament is defined as
$$\frac{\text{force acting on the thread}}{\text{cross-sectional area of the thread}}.$$

The radius of the filament is $(4.5 \pm 0.1)\,\mu\text{m}$.

(i) Determine the absolute uncertainty in the cross-sectional area of the filament.

(ii) The filament breaks when the force acting on it is 85 mN. The stress at which steel breaks is $1.0\,\text{GN m}^{-2}$.

Deduce whether the filament can withstand more stress than steel.

c) Determine the energy required to stretch the filament from an extension of 20 mm to 50 mm.

Problem 2

A student collects data for a system in which a mass oscillates about a given position. The graph of variation of time period T with mass m is shown.

The uncertainty in T is shown and the uncertainty in m is negligible.

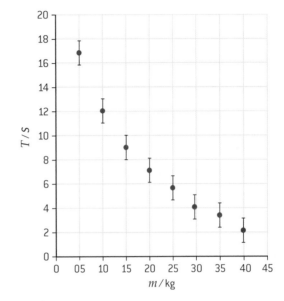

a) Draw the best-fit line for these data.

b) Determine the percentage uncertainty in T when $m = 10$ kg.

c) The student suggests that $T \propto \sqrt{m}$ and plots a graph of T against $\sqrt{m}$.

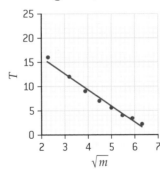

(i) Calculate the gradient of the graph.

(ii) Determine the intercept when $m = 0$.

d) Suggest the graph that the student should construct to confirm that $n = 2$ in the equation $T^n \propto m$.

A RELATIVITY

A.1 BEGINNINGS OF RELATIVITY

You must know:

✔ the definition of a reference frame

✔ what is meant by Galilean relativity

✔ Newton's postulates regarding space and time

✔ how the description of forces acting on a stationary charge or a moving charge depend on the reference frame

✔ Maxwell's postulate regarding the constancy of the speed of light in a vacuum.

You should be able to:

✔ use the Galilean transformation equations

✔ determine whether the force on a moving charged particle is electric or magnetic for a specified reference frame

✔ determine the nature of an electromagnetic field as observed from different reference frames.

> An **inertial reference frame** is one in which Newton's first law of motion holds. No external force must act on an object in the frame. By implication, the frame is not accelerating.

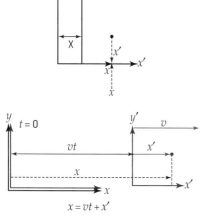

▲ **Figure A.1.1.** Two frames of reference connected by the Galilean transformation

A reference frame defines the motion of an object relative to others. Reference frames consist of an origin together with a set of axes.

An *inertial reference frame* is an extension of the reference frame idea.

Albert Einstein was not the first scientist to discuss reference frames; this was probably Galileo Galilei. He described a windowless cabin in a moving ship and suggested that butterflies in the cabin would always be observed to fly at random. An observer in the cabin could not deduce the motion of the ship by watching the butterflies.

Galileo's ideas suggest that:

• direction is relative

• position is relative

• stationary is not an absolute condition.

These lead to a set of equations (quoted for the x-direction only).

When the origin of a reference frame A' is a distance X from the origin of another reference frame A (Figure A.1.1), and the position of an object is x relative to the origin in frame A, the position of the object in frame A' is $x' = x - X$.

When the origin of an inertial frame A' is moving relative to the original of inertial frame A with speed v (and assuming that the origins coincide at time $t = 0$), the speed u of an object in frame A is related to the speed u' of the same object in frame A' by $u' = u - v$. Combining distances and speeds leads to $x' = x - vt$ at time t after the coincidence of the origins. These equations are known as the *Galilean transformations*.

Newton took Galileo's ideas and expressed them as the two *Newtonian postulates*.

Example A.1.1

Two identical spaceships, A and B, move along the same straight line.

A is moving away from the Moon at speed $0.40c$ and B is moving away from A at speed $0.30c$ relative to A.

Calculate the velocity of B relative to the Moon, using the Galilean transformation.

Solution

$u' = u + v = 0.30c + 0.40c = 0.70c$

All observers in inertial frames make the same deductions about physical law. James Clerk Maxwell connected electrostatics (Topic 5.1) and electromagnetism (Topics 11.1 and 11.2) by incorporating the speed of light in a vacuum. It was clear that, if observers in different inertial frames are to agree about physical laws, then they must all observe identical values for the speed of light in a vacuum. However, this is not what the Galilean transformation predicts. The equation $u' = u - v$ suggests that, if one observer moves at a constant velocity $(u - u')$ relative to another, then their observations of c will differ by v.

Transfer between inertial frames leads to unexpected changes in the descriptions of moving charge.

Two identical positive charges $+q$ are moving parallel at the same speed v. Figure A.1.2 gives two views of the same event:

(a) is the view from the inertial frame associated with an observer X

(b) is the view from the inertial frame of an observer Y. X moves with the charges and describes the effect between them as purely electrostatic in origin; they are repelled. Y, moving at a different speed, describes the effects differently, observing that the charges have a current associated with them.

Parallel currents in the same direction have a magnetic attraction between them. However, the two observations lead to the same physics (a repulsion of the two charges with the same force for both observers) as predicted by both the Newtonian and Einstein postulates.

Now consider a charge $+q$ moving parallel to a metal wire at a constant speed v. In the wire, the electrons are also moving at speed v. Protons in the wire are stationary. X is at rest relative to the protons and the wire. Y is moving with the positive charge: the moving positive charge is in the inertial frame of Y.

X sees the force on the positive charge as magnetic in origin. The moving electrons in the wire produce a magnetic field that is circular and centred on the wire. The charge is moving at 90° to this and experiences a repulsive magnetic force outwards from the wire.

Y does not agree with this. Y moves with the positive charge and with the electrons so the protons in the wire are moving to the left as far as Y is concerned. The proton separations in the wire are relativistically contracted for Y and have a higher linear density than that of the electrons. The positive protons feel a greater repulsive force than the attractive force of the electrons. Once again, the positive charge is repelled from the wire but Y interprets the force acting as electrostatic in origin. Once again, X and Y agree on the physical effects.

(a)

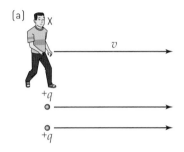

(b)

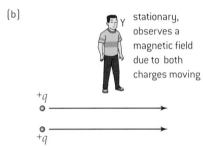

stationary, observes a magnetic field due to both charges moving

▲ Figure A.1.2. Two point charges moving in parallel

(a)

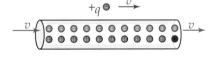

(b)

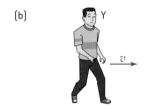

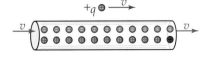

▲ Figure A.1.3. Moving charges in inertial reference frames.

⟩⟩ Assessment tip

You should be able to interpret this result with any combination of positive or negative moving charges and electron motion. Consider each possible case and ensure that you can explain it properly.

▲ The answer identifies correctly that the protons as viewed by Y will demonstrate only an electric (electrostatic) force.

▼ There is no mention that the Y force is repulsive. There is no discussion of the size of the repulsive force as observed by X; the need for a comparison is clearly flagged up in the question. X measures a smaller repulsive force than Y because there is an additional attractive magnetic force.

SAMPLE STUDENT ANSWER

Two protons are moving with the same velocity in a particle accelerator.

Observer X is at rest relative to the accelerator. Observer Y is at rest relative to the protons. Explain the nature of the force between the protons as observed by observer X and observer Y. [4]

This answer could have achieved 1/4 marks:

Observer X is moving relative to the reference frame. Therefore the protons would experience not only an electric force, but also a magnetic force. Since observer Y is at rest relative to the protons, they would only experience an electric force, since a proton only becomes magnetic once it is moving. If $v = 0$, no magnetic force will be experienced as seen by the equation $F = qvB$.

A.2   LORENTZ TRANSFORMATIONS

You must know:

✔ the two postulates of special relativity

✔ how two clocks can be synchronised

✔ the Lorentz transformations

✔ how to add velocities under special relativity

✔ what is meant by an invariant quantity

✔ the meaning of proper time and proper length

✔ the meaning of rest mass

✔ the meaning of spacetime interval

✔ the details and explanation of the muon decay experiment

✔ the meaning of time dilation and length contraction.

You should be able to:

✔ solve problems involving velocity addition

✔ use the Lorentz transformations to:

    ✔ describe how different measurements in spacetime by two observers can be converted into measurements in either frame of reference

    ✔ determine the position and time coordinates of various events

    ✔ show that two events that happen at different positions in space and observed to be simultaneous by one observer are not simultaneous in a different reference frame

✔ solve problems involving the time-dilation and length-contraction equations

✔ solve problems involving the muon-decay experiment.

In 1887, American scientists Albert Michelson and Edward Morley demonstrated that variations in light speed due to changes in measurement direction were so small that they fell below the precision of their experiment. They confirmed that the speed of light was independent of the reference frame in which it was measured.

This result led Einstein to modify the Newton postulates as follows.

• The laws of physics are the same in all inertial reference frames. (The same as Newton's first postulate.)

• The speed of light in free space (a vacuum) is the same in all inertial frames of reference. (Replacing Newton's concept of absolute space and absolute time.)

Lorentz introduced a gamma factor γ that compared the speed v of an inertial frame with c:

$$\gamma = \frac{1}{\sqrt{1 - \dfrac{v^2}{c^2}}}$$

He then modified the Galilean transformations to $x' = \gamma(x - vt)$ and $\Delta t' = \gamma \Delta t$. x refers to the position of a point so that the length of an object is the difference Δx between x_1 and x_2. This difference is the same in two inertial reference frames according to the Galilean transformation, but not according to Lorentz. In relativistic terms, when the difference Δx_1 in one frame is $(x_1 - x_2)$, then the difference Δx_2 in a second inertial frame moving relative to the first with γ is $\Delta x_2 = \gamma(\Delta x_1 - v\Delta t)$.

The important changes are that the difference in length now depends also on the speed of the second frame relative to the first, and that a factor depending on time also appears in the equation. Similarly, the time t' in the second inertial reference frame relative to the first becomes $t' = \gamma\left(t - \dfrac{v}{c^2}x\right)$.

Sometimes, the position and time are known in frame S' and the equivalent values in frame S are needed: in this case, the inverse Lorentz transformations are required.

Lorentz transformations and **inverse Lorentz transformations** (for transformation in the x-direction).

Galalilean	Lorentz	Inverse Lorentz
$x' = x - vt$	$x' = \gamma(x - vt)$ $\Delta x' = \gamma(\Delta x - v\Delta t)$	$x = \gamma(x' + vt')$
$t' = t$	$t' = \gamma\left(t - \dfrac{vx}{c^2}\right)$ $\Delta t' = \gamma\left(\Delta t - \dfrac{v\Delta x}{c^2}\right)$	$t = \gamma\left(t' + \dfrac{vx'}{c^2}\right)$

Frame A is moving relative to frame B with a constant velocity v.

An object in A is moving at u_A relative to A; an object in frame B is moving at u_B relative to B.

The speed of the object in A as measured by an observer in B is

$$u'_A = \frac{u_A + v}{1 + \dfrac{u_A v}{c^2}}.$$

The speed of the object in B as measured in B is $u'_B = \dfrac{u_B - v}{1 - \dfrac{u_B v}{c^2}}$.

These are given in the data booklet as $u' = \dfrac{u - v}{1 - \dfrac{uv}{c^2}}$

The transformations for length and time lead to expressions for the relative velocities when one or both objects are moving at speeds close to that of light. The equations follow from the usual definition of speed from Topic 2.1 but using the Lorentz transformations.

Example A.2.1

Two rockets, A and B, move along the same straight line when viewed from Earth.

A is travelling away from Earth at speed $0.80c$ relative to Earth.

B is travelling away from A at speed $0.60c$ relative to A.

a) Calculate the velocity of B relative to the Earth according to Galilean relativity.

b) Calculate the velocity of B relative to the Earth according to the theory of special relativity.

c) Comment on your answers.

Solution

a) Using $u' = u - v$ with appropriate signs gives $0.60c + 0.80c = 1.40c$

b) Using the equation as given in the data booklet:

$$u' = \frac{u - v}{1 - \frac{uv}{c^2}} = \frac{0.60c + 0.80c}{1 + \frac{0.60c \times 0.80c}{c^2}} = 0.95c$$

c) In part a), the speed exceeds c and this is not possible under the Einstein postulates. Galilean relativity is not valid (except as an approximation at small speeds $v < c$).

The need to use appropriate transformations does not, however, mean that all quantities in different inertial reference frames differ. Some quantities are *invariant* (unchanged) between frames.

The invariant quantities are: **Proper time interval** Δt_0 (often shortened to proper time) is the time interval between two events occurring at the same place in an inertial reference frame. It is the shortest possible time that can be observed between two events. In any other reference frame, the time between two events is **dilated** (longer) and the time in this new frame is given by $\Delta t = \gamma \Delta t_0$.

Proper length L_0 is the length of an object as measured by an observer at rest relative to the object. Any other observation of the object will result in a length measurement L that is shorter (**contracted**) than the proper length: $L = \dfrac{L_0}{\gamma}$.

Rest mass m_0 is the mass of an object in the frame at which it is at rest.

Space time interval $\Delta s^2 = (c\Delta t)^2 - (\Delta x)^2 - (\Delta y)^2 - (\Delta z)^2$.

Note: in the DP physics course, we assume one-dimensional motion (in the x-direction).

So, $s^2 = (ct)^2 - (x)^2$ ignoring the Δ symbols.

For any inertial frame, $\Delta s^2 = (ct)^2 - (x)^2 = (ct')^2 - (x')^2$.

Electric charge (see Topic A.4, HL only)

The muon decay experiment

This observation involves time dilation (or length contraction) for its explanation.

Muons are produced at the top of the atmosphere when high-energy cosmic rays interact with air molecules. The muons have a known probability of decay after formation and a known speed that is an appreciable fraction of the speed of light. It is possible to predict the fraction of the muons that will reach the ground. In fact, many more muons reach the ground than are predicted because the muons are decaying in a reference frame in which they are at rest. However, to us (the observers) in a frame moving relative to the muon frame at about $0.98c$, this muon decay rate is dilated and appears much longer than in the frame of the muons.

The muon observation can also be explained in terms of length contraction. The distance from the top of the atmosphere is measured by us to be a few tens of kilometres. However, to the muons, the distance is contracted as they are moving fast relative to our reference frame. We know how many muons are expected to decay for every metre they travel in our frame, but the distance as measured by observers in the muon frame is much shorter. Fewer muons are removed from the beam as it travels from the atmosphere to the surface.

Example A.2.2

Muons are particles with a proper lifetime of $2.2\,\mu s$. They move downwards to the surface at a speed of $0.98c$. Explain, with calculations, why this experiment provides evidence for relativity.

Solution
One of the conclusions of special relativity is that, to a moving observer, time is dilated.

Assuming Galilean relativity, muons must reach the ground in $6.7\,\mu s$. So, only $\frac{1}{8}$ of them should arrive undecayed.

Special relativity predicts that γ for this event is $\dfrac{1}{\sqrt{1-\left(\dfrac{v}{c}\right)^2}} = \dfrac{1}{\sqrt{1-0.98^2}} = 5$

To an observer in the frame of reference of the Earth, the mean lifetime of the muons should be $5 \times 2.2 = 11\,\mu s$. Therefore, only a small fraction should have decayed before they reach the ground. This is the observed outcome and it confirms the special relativity theory.

Measurements made in one reference frame are invalid in another frame unless we apply the Lorentz transformations. There is also the issue of measuring time in two inertial frames that are at rest relative to each other.

When two clocks are in the same inertial frame but are separated by $10\,m$, an observer near to one clock will notice that the other clock reads a difference of $3 \times 10^{-7}\,s$. This is not a simple problem because moving the clock closer (assuming that they are not in the same place) means accelerating the clock. During this motion, the clock is no longer in an inertial frame. The way to move a clock from one frame to another is to do so infinitely slowly or to use a third clock (also moving very slowly) and transfer one clock reading to the second clock.

SAMPLE STUDENT ANSWER

An electron is emitted from a nucleus with a speed of $0.975\,c$ as observed in a laboratory. The electron is detected at a distance of $0.800\,m$ from the emitting nucleus as measured in the laboratory.

a) For the reference frame of the electron, calculate the distance travelled by the detector. [2]

This answer could have achieved 2/2 marks:

$$\gamma = \frac{1}{\sqrt{1 - \dfrac{(0.95c)^2}{c^2}}} = 4.5$$

$$l = \frac{l_o}{\gamma} \qquad l = \frac{0.800\,m}{4.5} = 0.178\,m$$

b) For the reference frame of the laboratory, calculate the time taken for the electron to reach the detector after its emission from the nucleus. [2]

This answer could have achieved 2/2 marks:

$$v = \frac{d}{t} \quad t = \frac{d}{v} = \frac{0.800\,m}{0.975\,c} = 2.74 \times 10^{-9}\,s$$

c) For the reference frame of the electron, calculate the time between its emission at the nucleus and its detection. [2]

This answer could have achieved 2/2 marks:

$$\Delta t = \gamma \Delta t_o \quad \Delta t_o = \frac{\Delta t}{\gamma} = \frac{2.74 \times 10^{-9}\,s}{4.50} = 6.08 \times 10^{-10}\,s$$

d) Outline why the answer to c) represents a proper time interval. [1]

This answer could have achieved 1/1 marks:

Because it is the time interval between two observers in the same point in space.

▲ This sequence of answers, a)–d), scores full marks. The solutions in a)–c) are clear and easy to follow. The written response to d) is an acceptable alternative definition for *proper time interval* to the one given on page 150. It would also be acceptable to say that it is the shortest time interval that it is possible to observe.

A.3   SPACETIME DIAGRAMS

You must know:

✓ what is meant by a spacetime diagram and how to solve problems involving simultaneity and kinematics using spacetime diagrams

✓ what is meant by a worldline and how to represent a moving object by its worldline which may be a straight line (constant velocity) or a curve

✓ what is meant by the twin paradox

✓ and how to represent more than one inertial frame of reference on a spacetime diagram.

You should be able to:

✓ represent an event on a spacetime diagram as a point

✓ represent time dilation and length contraction using a spacetime diagram

✓ determine, for a specific speed, the angle between a worldline on a spacetime diagram and the time axis

✓ describe and resolve the twin paradox using a spacetime diagram.

🔗 The DP physics course only requires you to use spacetime diagrams for one-dimensional motion.

A *spacetime diagram* (also known as a Minkowski diagram) allows a simple visualisation of events in one or more reference frames. Explanations of relativistic phenomena are facilitated using the diagrams too.

Figure A.3.1 shows a spacetime diagram (with a *ct-x*-axis).

Particle A is stationary in this *ct-x* reference frame and is situated −1.0 m from the origin of the diagram. The single line parallel to the time axis is called the *worldline* of this particle. Because it is stationary in this frame, the worldline is parallel to the time axis.

Particle B is moving at a constant velocity of 0.5c. At $t = 0$ it was at the origin of the reference frame of the spacetime diagram. The worldline of this particle is a straight line moving through the spacetime diagram. The speed is 0.5c because, when the distance from the origin is 4 km, the time is $ct = 8.0$ and therefore $t = \dfrac{8.0}{c}$ and $v = \dfrac{4.0}{\left(\dfrac{8.0}{c}\right)} = \dfrac{4}{8}c$.

Particle C is accelerating from rest. It began at the point $(-2\,\text{km}, 0)$; the worldline for acceleration is a curve.

The angle θ between the B worldline and the ct-axis gives the speed of B: $\tan\theta = \dfrac{\text{opposite}}{\text{adjacent}} = \dfrac{X}{cT} = \dfrac{v}{c}$ so angle $\theta = \tan^{-1}\left(\dfrac{v}{c}\right)$.

For Figure A.3.1, θ is measured to be 27° and $\tan 27 = 0.5$ so $v = 0.5c$.

Particle B from Figure A.3.1 can also be displayed on a spacetime diagram drawn for the particle A reference frame (Figure A.3.2).

B is moving at 0.5c relative to A and is at the A origin (coincident with the B origin) at $t = 0$ in both frames. Therefore, as far as A is concerned, the worldline for B is the same as the time axis for B because B is stationary in its own frame.

The B time and distance axes both swing around in the A diagram, as shown in Figure A.3.2. The B axes are labelled as ct' and x'. This is a particularly helpful way to compare two inertial reference frames as it makes paradoxes in simultaneity easier to resolve.

A final element of Figure A.3.2 is a light cone which has been drawn from the event E that occurs at a position $x = 1.0\,\text{km}$ and a time $ct = 5.0\,\text{km}$. This shows the path of light in the reference frame of A. As nothing can exceed the speed of light, all ct'-axes must lie between the A ct-axis and the light cone.

The intersection of the two sets of axes give us important information about the way A and B perceive the timing of E in their reference frames.

- W is the time ($ct = 5$) at which E happens in the reference frame of A.

- X (the intersection of the light cone with ct) is the time when an observer on the worldline through the origin in A *sees* E happen.

- Y (the intersection of the light cone with ct') is the time when an observer on the world line of ct' sees E happen.

- Z (the intersection of E with a line constructed through E and parallel to x') is when E happens in frame B.

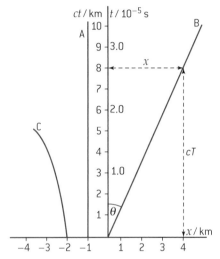

▲ Figure A.3.1. Spacetime diagram in the reference frame of A

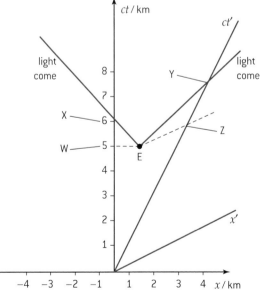

▲ Figure A.3.2. Events and a light cone on a spacetime diagram

Observers in the two frames do not agree about the timing of events. This is a consequence of the constancy of the speed of light to all observers.

Relativistic simultaneity is often explored through the use of paradox. A classic example is that of a train moving at high speed towards a tunnel.

Example A.3.1

A train of proper length 100 m is moving to the right towards a tunnel of proper length 80 m that has doors at each end. The speed of the train is 0.8c.

Explain why, in the tunnel reference frame, the doors can be shut on the train but, in the train frame of reference, the tunnel is too short for this to happen.

Solution

γ for the train is 1.67, so in the rest frame of the tunnel the length of the train is 60 m.

It will obviously fit into the tunnel in the tunnel reference frame. The doors are shut when the train is in the centre of the tunnel (with 20 m space in front of it).

So, at a time of $\dfrac{20}{0.8c} = \dfrac{20}{0.8 \times 3 \times 10^8} = 83$ ns after the doors shut, the train hits the doors.

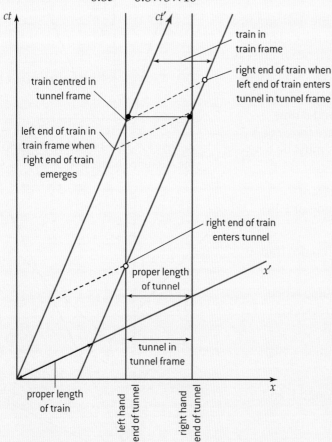

▲ Figure A.3.3. Spacetime diagram for the train–tunnel simultaneity problem

From the train frame of reference, the tunnel is contracted and approaches the train (moving to the left). The contracted length of the tunnel is 40 m and so the train is too long for the tunnel.

However, both doors can still shut for an instant. This is because the clocks (observers) in the tunnel and on the train that measure the timing of the doors shutting cannot ever agree.

Suppose there is a pair of clocks at each end of the tunnel. These clocks are synchronized in the tunnel frame. In the train frame, the tunnel clocks disagree. The one at the right-hand end of the tunnel is ahead of the left-hand clock by an amount of

$$\frac{vL}{c^2} = \frac{0.8 \times 80}{3 \times 10^8} = 210 \text{ ns}.$$

In the time interval between the right-hand clock and the left-hand clock reading zero according to the train observers, the left-hand door of the tunnel will have moved past the left-hand end of the train.

This problem is made clear in a spacetime diagram (Figure A.3.3). The tunnel frame is ct-x, the train frame is ct'-x'.

The twin paradox

Mark, one of a pair of twins, uses a spaceship to leave Earth and travel to a distant star at a constant speed with Lorentz factor γ.

On arrival at the star, he travels back to Earth at the same speed to find that his twin, Maria, has aged by $2T$ but Mark has only aged by $\dfrac{2T}{\gamma}$.

This is what we would expect from time dilation. However, a paradox arises because in Mark's frame, Maria has moved away from **him** at γ.

So, why is Maria not the younger twin on Mark's return?

Initially, the two reference frames in the paradox appear symmetrical. In fact, they are not. Maria has remained in an inertial (non-accelerated) frame. Mark, on the other hand, was accelerated at the beginning of the journey and decelerated at the end. To turn around at the star he also needed to accelerate. Moving out of the inertial frame breaks the symmetry and explains the paradox.

This can be seen on a spacetime diagram (Figure A.3.4) drawn in Maria's frame.

Maria remains on the worldline that is coincident with her ct axis. But Mark moves on worldline ct' out to the star.

Maria thinks that Mark arrives at the star at Q. Mark thinks that Maria observes his arrival time as R—they no longer agree on simultaneous events.

To avoid the acceleration–deceleration issue at the star, we imagine that Jay comes from the star at Mark's original speed towards Earth and that Mark synchronizes a clock with Jay as they pass.

Jay's worldline returning to Maria is shown. Jay thinks that Maria is at S as he leaves. Maria's ageing happens (so far as Jay and Mark are concerned) between Q and S.

> **>> Assessment tip**
>
> Study the annotations on the diagram carefully.

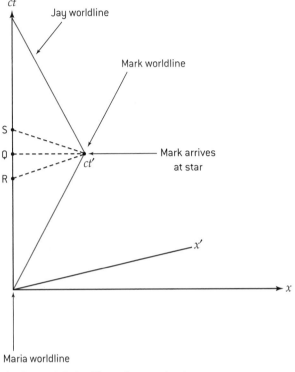

▲ Figure A.3.4. The twin paradox in a spacetime diagram

Example A.3.2

Four light beacons are used to guide a rocket A to a docking station. The events P, S, R and Q are flashes from the four beacons and are shown in the spacetime diagram.

The diagram shows the reference frames of Earth $(ct\text{-}x)$ and A $(ct'\text{-}x')$.

Deduce the order in which:

a) the beacons flash in the reference frame of A

b) an Earth observer sees the beacon flashing.

Solution

a) Construct lines through the events parallel to x'. These show that P and Q occur at the same time and that R then flashes before S.

b) Construct light cones from the events. These cross the ct-axis (and are seen on Earth) in the order P, R, Q and S simultaneously.

An observer on Earth watches a rocket A. The spacetime diagram shows part of the motion of A in the reference frame of the Earth observer. Three flashing light beacons, X, Y and Z, are used to guide rocket A. The flash events are shown on the spacetime diagram. The diagram shows the axes for the reference frames of Earth and of rocket A. The Earth observer is at the origin.

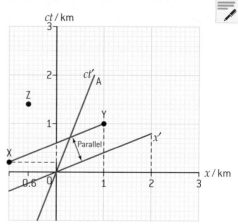

a) Using the graph above, deduce the order in which the beacons **flash** in the reference frame of the rocket.   [2]

This answer could have achieved 2/2 marks:

> ▲ This is a complete answer that shows a good understanding of the implications of spacetime diagrams. The diagram is well done and the student has taken time to indicate (with the word *parallel*) the essentials of the construction.

Beacons X and Y would happen at the same time before Z, because the worldline is parallel to the x' axis, meaning that in the reference frame of A, they would happen at the same time. Z would flash later, since it is further up the ct' axis than X and Y.

b) Using the graph above, deduce the order in which the Earth observer sees the beacons flash.   [2]

This answer could have achieved 2/2 marks:

> ▲ This is correct and gains full marks...

> ▼ ...however, the student could have made this clearer on the diagram. Light cones drawn from Z and Y intersect the worldline at $ct = 2$ km meaning that the events Y and Z are seen simultaneously.

Implying all beacons flash light going at the same velocity, X will be seen first by the observer on the Earth. Y and Z would arrive simultaneously, because Y flashes 0.4 km before, but has to travel 0.4 km, hence they, Y and Z, arrive at the same time after X.

A.4   RELATIVISTIC MECHANICS (AHL)

You must know:

✔ what is meant by rest energy and total energy

✔ what is meant by relativistic momentum

✔ what is meant by particle acceleration

✔ the implications of relativistic mechanics for photons

✔ that electric charge is an invariant quantity

✔ that mass and momentum can be expressed in units of MeV c^{-2} and MeV c^{-1} respectively.

You should be able to:

✔ describe the conservation laws of energy and momentum in the context of special relativity

✔ determine the potential difference necessary to accelerate a particle to a given speed or energy

✔ solve problems involving momentum and relativistic energy in collisions and particle decays

✔ calculate the wavelength of photons emitted during a relativistic decay.

Einstein needed to modify physical ideas of energy and momentum for their conservation laws to hold in all inertial frames. The rest mass m_0 of a particle—the mass as observed in a frame in which the particle is at rest—is invariant. Einstein showed that there was an equivalence between energy and mass expressed as $\Delta E = c^2 \Delta m$, and for the rest energy E_0 this is $E_0 = m_0 c^2$.

The equation $E^2 = p^2 c^2 + m_0^2 c^4$ can be rewritten as $(m_0 c^2)^2 = E^2 - (pc)^2$.

The quantity on the left-hand side of the equation is (rest mass)^2 which you know is invariant. The right-hand side must also be invariant and, for any two inertial frames, $E^2 - (pc)^2 = E'^2 - (p'c)^2$.

Placing a charged particle in an electric field causes it to accelerate. The kinetic energy ΔE_{ke} transferred to a particle of charge q when accelerated through a potential difference V is $\Delta E_{ke} = qV$. Electric charge is an invariant quantity.

When dealing with energy transfers involving particles, the units MeV c^{-1} for mass and MeV c^2 for energy are often used (instead of kilogramme and joule). This is the same as setting c to be 1 and using the equation $E^2 = p^2 + m_0^2$.

> The **total energy of a particle** is $E = E_0 + E_K$ (rest mass + kinetic energy E_K, ignoring potential energy) and $E = m_0 \gamma c^2$.
>
> So, $E_K = m_0 (1 - \gamma) c^2$.
>
> Momentum p is conserved in special relativity provided γ is incorporated into the equations.
>
> So $p = \gamma m_0 v$ and $E^2 = p^2 c^2 + m_0^2 c^4$.

Example A.4.1

Two protons are approaching each other head-on. The speed of each proton is $0.50c$ as measured by an observer in a laboratory.

a) Calculate the relative speed of approach as measured in the reference frame of one proton.

b) Determine the potential difference through which a proton must be accelerated in order to reach a speed of $0.70c$.

Solution

a) $u' = \dfrac{u + v}{1 + \dfrac{uv}{c^2}} = \dfrac{c}{1 + \dfrac{0.25c^2}{c^2}} = \dfrac{4}{5} = 0.80c$

b) The equation for the kinetic energy gain of the proton is
$E_{ke} = m_0 (1 - \gamma) c^2$

$\gamma = 1.67$ and $E_k = 1.67 \times 10^{-27} \times 0.67 \times 9 \times 10^{16} = 1.0 \times 10^{-10}$ J
$\qquad = 1.6 \times 10^{-19} \times V$

The pd required is 630 MV.

Example A.4.2

a) Calculate the potential difference through which a proton, starting from rest, must be accelerated for its mass–energy to be equal to four times its rest mass–energy.

b) Calculate the momentum of the proton after acceleration.

Solution

a) Kinetic energy gain $= 3m_p = 3 \times 938c^2 = 2810$ MV

b) Using $E^2 = p^2 c^2 + m_0^2 c^4$:

$16m_p^2 c^4 = p^2 c^2 + m_p^2 c^4$

So $p^2 c^2 = 15 m_p^2 c^4$

So $p^2 = 15 m_p^2 c^2 \Rightarrow p = \sqrt{15 m_p^2 c^2}$ or 3630 MeV c^{-1}.

> Topics 7.1 and 12.1 described photons as particles with energy but no mass. This implies that the equation for their energy is $E^2 = p^2 c^2$. Momentum can be assigned to a photon as
>
> $p = \sqrt{\dfrac{E^2}{c^2}} = \dfrac{E}{c}$ and therefore $p = \dfrac{h}{\lambda}$,
>
> where h is Plank's constant.

A lambda Λ^0 particle at rest decays into a proton p and a pion π^- according to the reaction:

$$\Lambda^0 \rightarrow p + \pi^-$$

where the rest energy of p $= 938$ MeV and the rest energy of $\pi^- = 140$ MeV.

The speed of the pion after the decay is $0.579c$. For this speed $\gamma = 1.2265$. Calculate the speed of the proton. [4]

This answer could have achieved 3/4 marks:

▲ The answer begins well. Although not mentioned, the conservation of momentum has been used to recognize that the particles after the decay have equal (and opposite) momenta. The solution is completely correct up to the calculation of the proton speed.

momentum for $\pi^- = 1.2265 \times 140 \times 0.579 = 99.42009 =$ momentum of p

proton $= 99.42009 \div 938 = 0.105992 = \gamma v$

$\therefore$ v for proton $= 0.105992 \div 1.2265 = 0.086418c = $ ans

▼ At this point, the solution is incorrect: the speed of the proton does not equal the speed of the pion as they have equal magnitude of momentum but different masses. The value for γ is not 1.2265 and must be incorporated as

$$\frac{v}{\sqrt{1 - \dfrac{v^2}{c^2}}} = 0.105992c.$$ The final

correct answer is $0.105c$.

A.5 GENERAL RELATIVITY (AHL)

You must know:

✔ the equivalence principle

✔ why light paths are bent in the presence of mass

✔ the definition of gravitational redshift

✔ details of the Pound–Rebka–Snider experiment

✔ the definition of Schwarzschild black holes

✔ the definition of an event horizon

✔ why time dilates near a black hole

✔ applications of the general theory of relativity to the Universe as a whole.

You should be able to:

✔ use the equivalence principle to deduce and explain the bending of light near masses and gravitational time dilation

✔ calculate gravitational frequency shifts

✔ describe an experiment to observe and measure gravitational redshift

✔ calculate the Schwarzschild radius of a black hole

✔ apply the equation for gravitational time dilation near the event horizon of a black hole.

The Einstein principle of equivalence states that gravitational effects cannot be distinguished from inertial effects. The principle can be illustrated using a thought experiment involving two observers watching the same events as one of them releases an object with mass.

Situation ①

Observer X is in an elevator (lift) that has no windows. Observer Y is outside the elevator and not connected to it; Y can see what happens inside.

The experiment is carried out well away from any masses or gravitational fields. The elevator is moving at constant velocity when X releases the object. It stays where it is, as no forces act on it. Both observers agree about this.

X repeats the experiment but, this time, as the object is released, the elevator begins to accelerate in the direction of the elevator ceiling. X describes the motion of the object as accelerating downwards. Y does not agree; to Y, the object stays stationary and the elevator and X accelerate around it.

Situation ②

When the elevator is at rest on the Earth's surface, X releases the object and interprets the motion of the object exactly as before (acceleration downwards). Y, on the other hand, will interpret it as a gravitational effect.

X cannot distinguish between situation ① and situation ②. This is the principle of equivalence.

The general theory suggests that there is no absolute motion, only relative motion. The status of an inertial frame of reference does not hold in the general theory. All observers, under the general theory, have the same status and obey the same laws.

The principle of equivalence also leads to some conclusions about the influence of gravitational fields on light. Imagine (Figure A.5.1) the same elevator above the Earth's surface, this time with a hole in one side. A pulse of light is shone through the hole. The light travels in a straight line and passes through another hole on the opposite side of the elevator. Both observers agree on this.

Another time as the light pulse enters the elevator, the elevator starts to accelerate downwards in the Earth's gravity field. The light enters the elevator and the observer in the elevator still sees the light exit through the hole because both the light and the observer are subject to the same gravitational field.

Observers X and Y must draw the same physical conclusions about the experiment. So observer Y must also see the light arrive through one hole and leave through the other. The only way these views can be reconciled is when Y describes the light as moving in a curved path as it travels in a gravitational field. Light, like any object, moves in a curved path in spacetime that has been warped by the presence of a nearby mass.

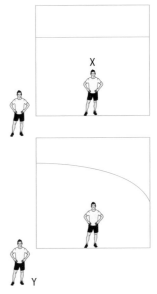

▲ Figure A.5.1. Principle of equivalence for light

Example A.5.1

A boy stands on the floor of an elevator. He throws a ball in a direction initially parallel to the floor. The ball follows a downwards curved path to the floor.

Explain whether the boy can deduce that the elevator is at rest.

Solution
The boy cannot make this deduction. The equivalence principle states that gravitational and inertial effects cannot be distinguished. The alternative is that the elevator and the boy could be accelerating upwards; these will give identical outcomes.

Gravity, according to the general theory and the equivalence principle, affects the motion of light.

A spaceship contains an observer and two light sources. One light source F, is at the front of the ship near the observer and the other light source R, is at the rear. The ship is moving at constant velocity relative to the Universe with the velocity vector in the direction of RF.

At the instant when the spaceship begins to accelerate, both light sources begin to emit light of the same frequency. The light from R takes a time $\Delta t = \dfrac{\Delta h}{c}$ to arrive at the observer. The light from F arrives almost instantaneously, because F and the observer are close.

During the time the light takes to reach the observer from R, the speed of the spaceship and observer will have changed by $\Delta v = a\Delta t = a\dfrac{\Delta h}{c}$, where a is the acceleration. The observer measures the speed of the light from R as c, as usual, but experiences a Doppler shift because there is now a difference between the source speed and the observer speed. The observed frequency shift is $\dfrac{\Delta f}{f} = \dfrac{\Delta v}{c} = a\dfrac{\Delta h}{c^2}$.

The same problem can be considered in the context of the spaceship resting on the Earth's surface with R below F. A photon reaching the observer from R must have transferred energy to the gravitational field to reach the observer. This energy loss corresponds to a reduction in the frequency (because $E = hf$) and so there must be a redshift as the wavelength increases.

In developing his general theory of relativity, Einstein proposed several tests, including the gravitational redshift of light.

This effect was confirmed by American physicists Pound, Rebka and Snider in 1959. They fired a gamma ray beam upwards to a detector 22 m above the gamma source and repeated the experiment firing downwards.

The gamma photons should experience a fractional change in energy $\dfrac{\Delta E}{E} = 2g\dfrac{\Delta h}{c^2}$. The values that Pound and his co-workers measured for the fractional energy changes compared well with the theoretical change, which confirms this test of general relativity.

Gravitational time dilation: a redshift is equivalent to a clock at R appearing to tick more slowly to the observer near F. Similarly, an observer on top of a mountain thinks that time runs more slowly at sea level.

Example A.5.2

An observer is close to the Earth's surface.

Laser A directs a light beam horizontally towards the observer who measures the frequency to be 4.8×10^{14} Hz.

Another identical laser B, 150 m below the observer, fires a light beam vertically upwards towards the observer.

a) Calculate the frequency difference between the two lasers, A and B, as measured by the observer.

b) Explain which laser has the higher frequency to the observer.

c) State one assumption made in your calculation in part a).

Solution

a) $\Delta f = g \dfrac{\Delta h}{c^2} f = 9.81 \times \dfrac{150}{\left(3 \times 10^8\right)^2} \times 4.8 \times 10^{14} = 7.9$ Hz

b) Laser A has the higher frequency because the photon loses energy as it rises from B. This loss of energy is translated into a frequency shift as $\Delta E = h \Delta f$.

c) One assumption is that the value of g does not change over the vertical change of 150 m.

German physicist and astronomer Karl Schwarzschild provided one of the first exact solutions of the field equations that constitute Einstein's general theory. He was able to derive equations for the gravity field that surrounds a spherical non-rotating uncharged mass.

His solution leads to the redshift in wavelength for a photon in the gravitational field of the mass M. This fractional wavelength shift is $\dfrac{\Delta \lambda}{\lambda} = \dfrac{R_s}{2r} = \dfrac{GM}{rc^2}$, where r is the distance from the centre of the mass to the point where the photon is emitted and R_s is the *Schwarzschild radius*.

> **Schwarzschild radius,** $R_s = \dfrac{2GM}{c^2}$.
>
> This has the dimensions of length.

For values of $r > R_s$, gravity applies as normal, but inside the sphere of radius R_s, the normal structure of spacetime does not apply. When $r = R_s$, there is a transition between the two regimes. This distance from the centre of the mass is known as the *event horizon*.

Near the event horizon, spacetime is extremely warped due to the strong gravitational field. Mass collapses towards the centre of the black hole.

Clocks in the region of a strong gravitational field run more slowly than in the absence of gravity. This is true near an event horizon. As a clock moves towards the transition, external observers see it tick more and more slowly with the clock never quite crossing the event horizon. The light emitted by the clock is gravitationally redshifted as the clock approaches R_s. The clock (and any observer unfortunate enough to be travelling with it) will observe its own passage through the event horizon in a finite amount of proper time.

> Events inside the **event horizon** defined by R_s cannot influence observers outside it. It represents the surface where gravitational pull is so large that nothing can escape, not even light itself. The event horizon is the surface at which the speed needed to escape from the mass is equal to the speed of light – this is the origin of the term black hole.

For a non-rotating mass with a Schwarzschild radius R_s, the proper time interval Δt_0 is related to the time interval Δt measured by a distant observer at distance r from the centre of the mass by

$$\Delta t = \frac{\Delta t_0}{\sqrt{1 - \dfrac{R_s}{r}}}.$$

>> Assessment tip

When you write about spacetime in the vicinity of a black hole or event horizon, always give the impression of *extreme* warping. Remember that spacetime is always warped by the present of mass. In the unusual conditions around a black hole, the warping is much greater than normal.

▲ The crucial idea that light cannot escape is here.

▼ However, it would have been better to have seen a reference to spacetime warping rather than the idea that gravitational force is too strong.

▼ The terms in the equation are not referenced in the answer or used; 5.0 s is not mentioned as a proper time for the event on S-2. The phrases 'dilated more' and 'dilated less' beg the question, 'More or less than what?' The points to be made here are:

• the proper time for the event on S-2 is 5.0 s, so any times observed in a different frame must, by general relativity, *always* be greater than 5.0 s

• when S-2 is closer to the black hole, the observed time period will be dilated more than when it is at its extreme distance from the black hole.

Example A.5.3

a) Explain, with reference to the motion of light, what is meant by the Schwarzschild radius.

b) Deduce the distance from the event horizon to the centre of a star of mass 2×10^{30} kg.

Solution

a) The Schwarzschild radius is the largest distance from the centre of a mass to the point outside the mass at which photons of light cannot escape its gravitational field.

This is because mass curves spacetime and photons follow the shortest path between two points through curved spacetime.

Photons cannot escape in conditions where extreme warping of spacetime prevents photons from escaping.

b) $R_s = \dfrac{2GM}{c^2} = \dfrac{2 \times 6.7 \times 10^{-11} \times 2 \times 10^{30}}{\left(3 \times 10^8\right)^2} = 3.0$ km

---

SAMPLE STUDENT ANSWER

It is believed that a non-rotating supermassive black hole is likely to exist near the centre of our galaxy. This black hole has a mass equivalent to 3.6 million times that of the Sun.

a) Outline what is meant by the event horizon of a black hole. [1]

This answer could have achieved 1/1 marks:

The event horizon is an imaginary surface of a sphere where the gravitational pull is so strong that light cannot escape.

Star S-2 is in an elliptical orbit around a black hole. The distance of S-2 from the centre of the black hole varies between a few light-hours and several light-days. A periodic event on S-2 occurs every 5.0 s.

b) Discuss how the time for the periodic event as measured by an observer on the Earth changes with the orbital position of S-2. [2]

This answer could have achieved 0/2 marks:

When the distance between S-2 and the black hole decreases, time is dilated less as

$\Delta t = \dfrac{\Delta t_o}{\sqrt{1 - \dfrac{R_s}{r}}}$, and dilated more when the distance increases, to the periodic event changes.

Practice problems for Option A

Problem 1

A space station is at rest relative to Earth and carries clocks synchronized with clocks on Earth.

A spaceship passes Earth travelling at a constant velocity with $\gamma = 1.25$.

a) Calculate, in terms of c, the speed of the spaceship relative to Earth.

b) As the spaceship passes Earth, a radio signal is emitted from the Earth that is reflected by the spacestation and later observed on the spaceship.

Construct and annotate a spacetime diagram to show these events.

Problem 2

a) Outline what is meant by proper length.

b) A pion decays in a proper time of 46 ns. It is moving with a velocity of $0.95c$ relative to an observer. Calculate the decay time of the pion as measured by the observer.

Problem 3

An unstable particle A decays into a particle B and its antiparticle $\bar{B}$.

A is at rest relative to the laboratory when it decays.

The momentum of B relative to the laboratory is $7.4\,\text{GeV}\,c^{-1}$.

The rest mass of B and $\bar{B}$ is $1.8\,\text{GeV}\,c^{-2}$.

Deduce the rest mass of A.

Problem 4

An electron is observed in a laboratory to have a total energy of 2.30 MeV.

a) Show that the speed of the electron is about $0.98c$.

b) The electron is detected at a distance of 0.800 m from its source in the laboratory frame.

(i) Calculate the distance travelled by the detector in the electron frame.

(ii) Calculate the time taken for the electron to reach the detector from the source in the laboratory frame.

(iii) Calculate the time taken by the electron to move between its source and the detector in the electron frame.

(iv) Suggest which of your answers to (ii) and (iii) is a proper time interval.

Problem 5

A spaceship leaves Earth with a speed $0.80c$.

a) Draw a spacetime diagram for the Earth's frame including the motion of the spaceship.

b) Label your diagram with the angle between the worldline of the spaceship and that of the Earth.

Problem 6

An electron and proton with equal and opposite velocities annihilate to produce two photons of identical energies. The initial kinetic energy of the electron is 2.5 MeV.

a) Determine the speed of the electron.

b) Calculate the energy and momentum of one of the photons.

B ENGINEERING PHYSICS

B.1 RIGID BODIES AND ROTATIONAL DYNAMIC

You must know:

- ✔ the definition of torque
- ✔ the definition of moment of inertia
- ✔ the definition of rotational equilibrium and linear equilibrium
- ✔ that equations of rotational kinematics apply under conditions of constant angular acceleration
- ✔ that angular momentum is conserved
- ✔ that Newton's second law can be applied in a modified form to angular motion
- ✔ the distinction between an object that is rolling and one that is slipping.

You should be able to:

- ✔ calculate torque for single forces and for couples
- ✔ sketch and interpret graphs that show the variation with time of angular displacement, angular velocity and torque
- ✔ solve problems involving the rotational equivalent of Newton's second law
- ✔ solve problems involving objects in both rotational and translational equilibrium
- ✔ solve problems involving the rotational quantities moment of inertia, torque and angular acceleration as analogies to linear quantities
- ✔ solve problems that involve rolling without slipping.

Topic 2 covers linear mechanics and the interaction of objects that are treated as points; this part of Option B deals with objects that have shape and size. Many of the quantities in rotational mechanics have direct analogies in linear mechanics.

When an object rotates about an axis with no translational motion and is displaced through an angle θ in a time t, it has an **angular velocity**

$$\omega = \frac{\theta}{t}.$$

Remember from Topic 9.1 that
$$\omega = 2\pi f.$$

When the initial angular speed ω_i changes to a final angular speed ω_f in time t, **its angular acceleration**

$$\alpha = \frac{(\omega_f - \omega_i)}{t}.$$

Assessment tip

You should be confident using the equations from Topic 2.1. Learn the links between linear quantities and rotational quantities and the way in which they are used. This topic emphasizes these links.

You should also be familiar with the quantities described in Topic 6.1.

Here is the correspondence between quantities in linear and rotational motion:

Linear quantities	Rotational quantities
$v = u + at$	$\omega_f = \omega_i + \alpha t$
$v^2 = u^2 + 2as$	$\omega_f^2 = \omega_i^2 + 2\alpha\theta$
$s = ut + \frac{1}{2}at^2$	$\theta = \omega_i t + \frac{1}{2}\alpha t^2$
$s = \frac{(v+u)}{2t}$	$\theta = \frac{(\omega_f + \omega_i)}{2t}$

Example B.1.1

A laboratory centrifuge reaches its working angular speed of $1100 \, \text{rad s}^{-1}$ from rest in $4.2 \, \text{s}$. The moment of inertia of the centrifuge is $7.6 \times 10^{-4} \, \text{g m}^2$.

Calculate the angular acceleration of the system.

Solution

Angular acceleration $\alpha = \dfrac{(\omega_f - \omega_i)}{t} = \dfrac{(1100 - 0)}{4.2} = 260 \, \text{rad s}^{-2}$

The next step is to examine the way in which the rotational equivalent of force connects to angular acceleration.

The rotational equivalent of mass is *moment of inertia I*. Moments of inertia depend on the particular axes of rotation. For two very small spheres, with mass m, connected by a light rod length l about an axis through the centre of the rod, I is given by

$$I = m\left(\frac{l}{2}\right)^2 + m\left(\frac{l}{2}\right)^2 = \frac{1}{2}ml^2$$

However, when the rotational axis is changed to the centre of one of the spheres, I becomes $0 + ml^2$.

The moment of inertia for a bicycle wheel of mass m and radius r about an axis of rotation through the centre of the wheel is simply mr^2 (assuming that the spokes and the centre bearing are very light compared with the rim and tyre).

The rotational equivalent of linear force is *torque*. A torque exists when a force F acts at a distance r from an axis of rotation and produces a turning effect. Torque is a vector and, therefore, has a direction; this is defined by a right-hand rule (rather like Fleming's rule in electromagnetism was defined by a left-hand rule). The relationship is shown in Figure B.1.1.

The link between Newton's second law, $F = ma$, and

$$F = \frac{\text{change in momentum}}{\text{time taken}}$$ also applies to rotational motion when

angular momentum has been defined:

$$\Gamma = \frac{\text{change in angular momentum}}{\text{time taken}}.$$

Angular momentum of a system is conserved unless an external torque acts on the system. This has consequences when the moment of inertia of the system is modified with purely internal means. A common example of this is an ice skater spinning about a vertical axis through the centre of their body. They alter their rotational speed by pulling their arms into their body. This reduces their moment of inertia and they must speed up to conserve angular momentum. Notice that no external torque acts here because the movement of the arms can only give rise to an internal torque. The energy transfer involved must be produced by the skater.

Rotational kinetic energy links to linear kinetic energy. As usual, changes can be inelastic, elastic or have energy added depending on the changes in momentum of the system or systems involved.

Moment of inertia $I = \sum mr^2$, where m is the mass of an object and r is the distance from the axis for each part of the object.

The unit of moment of inertia is kg m^2.

>> **Assessment tip**

You will not be required to calculate moments of inertia. The value of I or the equation to compute it for a particular shape will be provided.

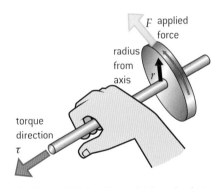

▲ **Figure B.1.1.** The right-hand rule to describe torque

Newton's laws of motion can be expressed in rotational contexts.

Newton's first law: Every rotating body continues to rotate at constant angular velocity unless an external torque acts on it.

Newton's second law: $F = ma$. Newton's second law in a rotational context is $\Gamma = I\alpha$.

Newton's third law: Action torque and reaction torque acting on a body are equal and opposite.

Angular momentum, L is defined as $I\omega$. The unit of angular momentum is kg m² s⁻¹.

The **rotational kinetic energy** $E_{K_{rot}}$ of a body with moment of inertia I and angular velocity ω is

$$E_{K_{rot}} = \frac{1}{2}I\omega^2$$

The change in rotational kinetic energy E_{Krot} of a body with moment of inertia I changing angular velocity between ω_i and ω_f is

$$\frac{1}{2}I\left(\omega_f^2 - \omega_i^2\right).$$

The unit of E_{Krot} is kg m² s⁻² which is the same as a joule.

Example B.1.2

A flywheel is accelerated from rest. The flywheel has a moment of inertia of 250 kg m² and takes 8.0 s to accelerate to 90 rev min⁻¹.

a) Calculate the angular acceleration of the flywheel.

b) Calculate the average accelerating torque acting on the flywheel.

c) Calculate the rotational kinetic energy stored in the flywheel at the end of its acceleration.

Solution

a) Angular speed of the flywheel $= \dfrac{90 \times 2\pi}{60} = 9.4\,\text{rad s}^{-1}$

and $\alpha = \dfrac{9.4}{8.0} = 1.18\,\text{rad s}^{-2}$

b) $\Gamma = I\alpha = 250 \times 1.18 = 295\,\text{N m}$

c) Rotational kinetic energy $= \dfrac{1}{2}I\left(\omega_f^2 - \omega_i^2\right) = \dfrac{1}{2} \times 250 \times 9.4^2$

$$= 1.1 \times 10^4\,\text{J}$$

When a cylinder **rolls** along horizontal ground, the point of contact between the cylinder and ground is at rest.

When a cylinder **slides** on the ground, the point of contact moves (slips) along the ground.

Because the contact point is at rest in rolling, the coefficient of static friction should be used in any calculations.

A cylinder of radius r rolls to the right at linear speed v. The top of the cylinder moves to the right at $v + r\omega$ and the bottom must be moving at $v - r\omega$. However, as $v - r\omega = 0$ (point at rest) then $v = r\omega$ and the top of the cylinder is moving with linear speed $2v$.

The total kinetic energy of a rolling object is $\dfrac{1}{2}I\omega^2 + \dfrac{1}{2}mv^2$. When this kinetic energy is gained by the object rolling down a slope with a vertical height change of h, $mgh = \dfrac{1}{2}I\omega^2 + \dfrac{1}{2}mv^2$.

A satellite approaches a rotating space probe at a negligibly small speed in order to link to it. The satellite does not rotate initially, but after the link they rotate at the same angular speed.

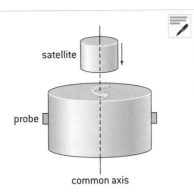

The initial angular speed of the probe is $16 \, \mathrm{rad \, s^{-1}}$.

The moment of inertia of the probe about the common axis is $1.44 \times 10^4 \, \mathrm{kg \, m^2}$.

The moment of inertia of the satellite about the common axis is $4.80 \times 10^3 \, \mathrm{kg \, m^2}$.

a) Determine the final angular speed of the probe–satellite system. [2]

This answer could have achieved 2/2 marks:

Angular momentum is conserved, $L = I.\omega \quad \omega s = \omega p_2$

$Ip_1 . \omega p_1 = Is.\omega s + Ip_2 . \omega p_2$

$Ip_1 . \omega p_1 = \omega p_2 (Is + Ip_2)$

$1.44 \times 10^4 \times 16 = \omega p_2 (1.44 \times 10^4 + 4.80 \times 10^3)$

$230400 = \omega p_2 \times 19200$

$\omega p_2 = 12 \, \mathrm{rad/s}$

▲ A well-presented solution that makes everything obvious to the examiner. The physics principles are stated and the substitutions are clear.

b) Calculate the loss of rotational kinetic energy due to the linking of the probe with the satellite. [3]

This answer could have achieved 3/3 marks:

$E_{krot1} = \dfrac{1}{2} . I_p . \omega_p^2 = \dfrac{1}{2} \times 1.44 \times 10^4 \times 16^2 = 1843200$

$E_{krot2} = \dfrac{1}{2} (I_p + I_s) \times \omega p_2^2 = \dfrac{1}{2} \times 19200 \times 12^2 = 1382400$

$E_{krot1} - E_{krot2} = 460800 \approx 460000 \mathrm{J} \; lost$

$460000 \mathrm{J}$ is lost.

▲ Again, the answer is well laid out and accurate. It is easy to award the full credit here. This is a model of how to answer a question.

B.2   THERMODYNAMICS

You must know:

✔ the first law of thermodynamics

✔ the second law of thermodynamics

✔ the definition of entropy

✔ the definition of isothermal and adiabatic processes

✔ that an isovolumetric process is carried out at constant volume

✔ that an isobaric process is carried out at constant pressure

✔ the definition of cyclic processes and that they can be visualized using pV diagrams, and how to sketch and interpret them

✔ what is meant by a Carnot cycle

✔ the definition of thermal efficiency and how to solve problems involving thermal efficiency.

You should be able to:

✔ describe the first law of thermodynamics as a statement of conservation of energy

✔ explain the sign conventions used in the first law of thermodynamics, $Q = W + \Delta U$

✔ solve problems involving the first law of thermodynamics

✔ describe the second law of thermodynamics using the Clausius interpretation and the Kelvin (Joule–Kelvin) interpretation

✔ describe the second law of thermodynamics in terms of, and as consequence of, entropy

✔ solve problems involving entropy changes and describe processes in terms of entropy change

✔ solve problems involving adiabatic changes for monatomic gases where $pV^{\frac{5}{3}}$.

🔗 As with Topic 3, Option B.2 covers the behaviour of gases, but takes a broader view. Here, we consider the general properties of systems in terms of changes they undergo and how these have an impact on the rest of the Universe.

A **thermodynamic system** defines the items of interest in a particular context. The system is separated from its **surroundings** by a **boundary**. The system together with the surroundings constitutes the **Universe**.

A pressure–volume (pV) diagram shows the changes in the pressure and volume of a gas as it moves between two or more states or around a closed cycle.

The first law of thermodynamics is an expression of the conservation of energy as it applies to systems; specifically, a system of a gas acted on by its surroundings. The law is written as $Q = W + \Delta U$. The terms in the equation when positive are:

- Q—the energy transferred **into** the system from the surroundings

- W—the work done **by** the system **on** the surroundings

- ΔU—the change in the internal energy of the system.

This equation can be applied to the four gas changes shown in Figure B.2.1. For all four changes, imagine that an ideal gas (system) is trapped inside a cylinder with a piston at one end (making up the boundary) with the surroundings being everything else in the Universe.

Some changes of state of a gas have special names that you should recognize.

Isobaric changes occur at constant pressure.

Isovolumetric changes occur at constant volume.

Isothermal changes occur at constant temperature.

Adiabatic changes occur without energy being transferred into or out of the gas.

Isobaric change

The work done by the system on the surroundings is at constant pressure. The energy transfer W when the piston of area A is moved through distance x is pAx. But, Ax is the change in volume ΔV, $W = p\Delta V$.

The first law becomes $Q = \Delta U + p\Delta V$.

The work done by or on the system is the equivalent of the area under the pV graph. This can be evaluated by either counting squares under the graph or by integration.

Isothermal change

There is no change to ΔU as this is the internal energy of the gas: $Q = W$.

Any thermal energy transferred into the system must appear as external work done by the system on the surroundings.

Isovolumetric change

The term W is zero because no work is done by or on the system (gas): $Q = \Delta U$.

Adiabatic change

No energy is transferred into or out of the system, and so $Q = 0$ implying that $\Delta U = -W$. Any external work done by the system must be at the expense of the internal energy. Put simply, when work is done on the surroundings, the temperature of the system must fall.

> ## ≫ Assessment tip
>
> When sketching pV graphs, make sure that the relative gradients of the isothermal and adiabatic changes are correct.
>
> Remember that the area underneath a pV graph is the energy transferred. This can be work done **on** the gas, or work done **by** the gas depending on the direction of the state change. When the gas expands, it is doing work; when it is compressed, work is done on it.
>
> $p \times V$ has the units of energy.

Example B.2.1

0.064 mol of an ideal gas is enclosed in a cylinder by a frictionless piston.

Two isotherms are shown on the pV diagram for 300 K and 500 K.

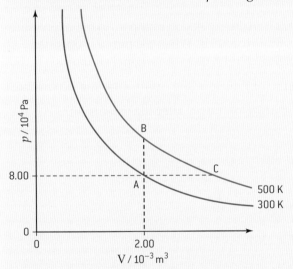

a) Explain how the first law of thermodynamics applies when the state of the gas is changed from:

   i) A to B at constant volume

   ii) A to C at constant pressure.

b) Calculate the heat energy absorbed by the gas in the change from:

   i) A to B

   ii) A to C.

Solution

a) The first law of thermodynamics is written as $Q = W + \Delta U$, where Q is the energy entering the gas from the surroundings, W is the work done by the system and ΔU is the change in the internal energy of the gas.

i) The change A to B is at constant volume, so $Q = \Delta U$ as no work (W) is done by the system on the gas. The temperature increases to reflect the change ΔU.

ii) The change A to C is at constant pressure and so the piston must move to allow this.

$$Q = W + \Delta U$$

In this case, the temperature increases *and* work is done in expansion.

b) i) $\Delta U = \dfrac{3}{2} nR(500 - 300) = 1.5 \times 0.064 \times 8.31 \times 200 = 160 \, \text{J}$

ii) $V_2 = V_1 \dfrac{T_2}{T_1} = 3.3 \times 10^{-3} \, \text{m}^3$

$p\Delta V = 8.0 \times 10^4 \times (3.3 - 2.0) \times 10^{-3} = 104 \, \text{J}$

So $\Delta Q = \Delta U + p\Delta V = 264 \, \text{J}$

When the state of an ideal monatomic gas changes from

(p_1, V_1) to (p_2, V_2) in an **adiabatic change**, $pV^{\frac{5}{3}} = \textbf{constant}$.

So $p_1 V_1^{\frac{5}{3}} = p_2 V_2^{\frac{5}{3}}$.

(The exponent is different when the gas has more than one atom in the molecule.)

This is why the gradient of an adiabatic change on a pV diagram is steeper for the same gas than when it undergoes an isothermal change (for which $pV = \text{constant}$).

Example B.2.2

An ideal monatomic gas is in an expansion pump at an initial pressure of 100 kPa and a temperature of 313 K. When the pump is operated, the gas expands adiabatically to 1.7 times its original volume.

a) Calculate the pressure of the air in the cavity after the expansion.

b) Calculate the temperature of the air after the expansion.

Solution

a) $pV^\gamma = \text{constant}$, where $\gamma = \frac{5}{3}$

$$p = 100 \times 10^3 \times \left(\frac{1}{1.7}\right)^{\frac{5}{3}} = 41.3 \text{ kPa}$$

b) $\dfrac{pV}{T} = \text{constant} \Rightarrow T = 220 \, \text{K}$

Carnot gave the first description of a theoretical heat engine—a device that operates through a cycle of gas state changes—known as the *Carnot cycle*. Energy is transferred into the working fluid (the gas in this case) at a high temperature, and energy is transferred out to the surroundings at a lower temperature. The remainder of the energy does work on the system.

Energy Q_1 is supplied to the gas trapped in a cylinder by a piston from a hot reservoir that is at a high temperature T_{hot}. The gas expands, and the piston will move until the pressure of the gas is the same as atmospheric pressure. The gas has done work under these conditions.

However, the gas is to work in a cycle, so it must now go back to its original state. This can only happen when an amount of energy Q_2 is rejected to a cold reservoir at a low temperature T_{cold}.

The thermal efficiency of the Carnot heat engine is given by:

$$\eta_{Carnot} = \frac{\text{useful work done}}{\text{energy input}} \equiv \frac{Q_1 - Q_2}{Q}$$

This assumes that all the energy $Q_1 - Q_2$ is transferred into useful work and there are no losses to friction, and so on (which is why Carnot's engine is only theoretical).

The cycle consists of two isothermal and two adiabatic changes (Figure B.2.1).

W to X: The gas is at T_{hot} and expands isothermally absorbing energy Q_1. Note that no energy goes into the gas because the change is isothermal.

X to Y: The gas expands adiabatically and the temperature falls to T_{cold}. The gas loses internal energy and continues to do work on the atmosphere—the piston moves as the gas expands.

Y to Z: The gas is now compressed isothermally to Z with no change to the gas internal energy. The work done on the gas is rejected as energy.

Z to W: The gas is compressed, again adiabatically, and all the work done on the gas increases its internal energy to return it to T_{hot}.

The net work done by the gas on the surroundings in one cycle is the area enclosed by the curve. This cycle is reversible and it can return to a previous energy state—this means that the cycle must be operated infinitely slowly—another reason why the Carnot cycle is theoretical.

> The thermal efficiency for the Carnot cycle can also be written as
> $$\eta_{Carnot} = \frac{T_{hot} - T_{cold}}{T_{hot}} = 1 - \frac{T_{cold}}{T_{hot}}.$$

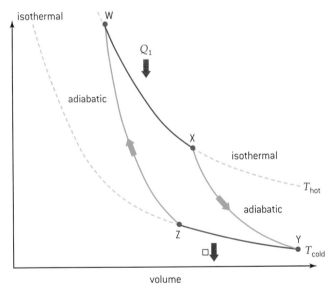

▲ **Figure B.2.1.** The Carnot cycle

Example B.2.3

This table shows some measurements made on an experimental heat engine.

temperature of heat source	830 °C
temperature of cooling system	17 °C
heat energy supplied per second	78 J
power output of heat engine	1.5 W

a) Calculate the maximum possible efficiency of the engine.

b) Suggest whether the actual efficiency of the heat engine approaches your answer to part a).

Solution

a) $\eta = \dfrac{T_{hot} - T_{cold}}{T_{hot}} = \dfrac{1103 - 290}{1103} = 74\%$

b) $\eta = \dfrac{P_{out}}{P_{in}} = \dfrac{15}{78} = 19\%$

The engine is significantly less efficient than the theoretical value.

The **Clausius statement** of the second law:

Energy cannot flow spontaneously from an object at a low temperature to an object at a higher temperature without external work being done on the system.

The **Kelvin statement** of the second law:

Energy cannot be extracted from a reservoir and transferred entirely into work.

The **entropy formulation** (due to Boltzmann) statement:

For any real process the entropy of the Universe must not decrease.

The change in **entropy** ΔS of the system is $\Delta S = \dfrac{\Delta Q}{T}$.

The unit of entropy is J K^{-1}.

The first law of thermodynamics equates work and energy transfer. The second law of thermodynamics lays out the situations in which energy can be transferred into work.

There are a number of ways to state the second law. Three of these are required in the DP physics course: the Clausius statement, the Kelvin statement and the entropy formulation statement.

Entropy is defined in terms of the energy ΔQ absorbed by a system and the kelvin temperature T at which the energy transfer occurs. Entropy is a measure of the disorder in a system. Any real process tends to increase disorder.

Consider a crystal of common salt (sodium chloride). When solid, the salt atoms are arranged regularly in a highly ordered way. Drop the crystal into water and it dissolves with many possible arrangements for the solution. The entropy of the system has increased. To restore the order to the crystal, the water must be evaporated either naturally or by heating. This process of decreasing the entropy of the dissolved salt to that of the solid again will cause other entropy increases in the Universe; the total of these changes always increases.

Example B.2.4

Calculate the entropy change when 1 kg of ice melts.

The specific latent heat of fusion of ice is 330 kJ kg^{-1}

Solution
The melting occurs at a temperature of 273 K.

The entropy change is $\dfrac{330 \times 10^3}{273} = 1.2$ kJ K^{-1}

SAMPLE STUDENT ANSWER

A heat engine operates on the cycle shown in the pressure–volume diagram. The cycle consists of an isothermal expansion AB, an isovolumetric change BC and an adiabatic compression CA. The volume at B is double the volume at A. The gas is an ideal monatomic gas.

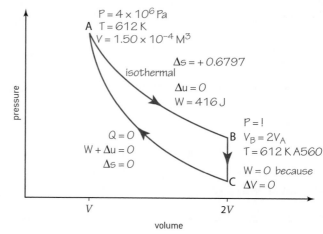

At A the pressure of the gas is 4.00×10^6 Pa, the temperature is 612 K and the volume is 1.50×10^{-4} m³. The work done by the gas during the isothermal expansion is 416 J.

a) i) Justify why the thermal energy supplied during the expansion AB is 416 J. [1]

This answer could have achieved 1/1 marks:

Because it is isothermal, $\Delta T = 0$, and $\Delta u = \dfrac{3}{2} nR\Delta T$

$\therefore \Delta u = 0$. $Q = \Delta u + W$, and if $\Delta u = 0$, $Q = W$.

In this case $W = 416 J$, so $Q = 416 J$.

▲ The answer correctly identifies that there is no change in U and therefore $Q = W$ so that Q is also 416 J.

The temperature of the gas at C is 386 K.

ii) Show that the thermal energy removed from the gas for the change BC is approximately 330 J. [2]

This answer could have achieved 2/2 marks:

$\Delta V = 0$, $W = 0$ so $Q = \Delta u$ $nR = \dfrac{PV}{T} = 0.980$ $\Delta u = \dfrac{3}{2} nR\Delta T$

$\Delta T = 386 - 612 = -226 K$

$\Delta u = \dfrac{3}{2} \times 0.98 \times (-226) = -332.227 = -330 J$

$Q = -330 J$, so 330 J is taken out of the gas.

▲ BC is at constant volume so no change in W for this part of the cycle; $Q = \Delta U$ and a calculation using $\dfrac{3}{2} nRT$ confirms the result. A clear answer.

iii) Determine the efficiency of the heat engine. [2]

This answer could have achieved 2/2 marks:

$e = \dfrac{W}{Q}$ $Q_{in} = 416 J$

useful work done $= 416 - \Delta u_{AC}$ $\Delta u_{AC} = -\Delta u_{BC} = +330$

$\therefore W = 416 - 330 = 86 J$

$e = \dfrac{W}{Q} = \dfrac{86}{416} = 0.207$

0.207 or 20.7%

▲ Once again, clear and well-presented work. It should be your aim to achieve this sort of quality in your examination answers.

b) State and explain at which point in the cycle ABCA the entropy of the gas is the largest. [3]

This answer could have achieved 3/3 marks:

B would have the highest entropy. Entropy difference is calculated wrong. $\Delta S \dfrac{\Delta Q}{\Delta T}$ and for the change AB

$\Delta S_{AB} = +0.680$. This means that the system gains entropy from A→B. A and C have the same entropy because $\Delta Q = 0$, as the change is adiabatic. Therefore, if the entropy at B is higher than at A, it will also be higher than that at C. Therefore B has the largest entropy.

▲ The answer is stated clearly and there is a chain of argument that supports the answer. Again, a model answer.

B.3   FLUIDS AND FLUID DYNAMICS (AHL)

You must know:

✔ the definitions of density and pressure

✔ Archimedes, principle and why objects are buoyant in a liquid

✔ what is meant by an ideal fluid

✔ the meaning of hydrostatic equilibrium

✔ Pascal's principle

✔ the significance of streamlines

✔ the definitions of laminar and turbulent flow

✔ the continuity principle, Bernoulli's equation and Stokes's law.

You should be able to:

✔ determine the force acting on buoyant objects using Archimedes' principle

✔ solve problems involving pressure, density, Pascal's principle, the Bernoulli equation, the continuity equation and Stokes's law

✔ explain effects in fluid flow using the Bernoulli effect

✔ describe the frictional drag that is exerted on small spheres during laminar flow

✔ define and determine the Reynolds number.

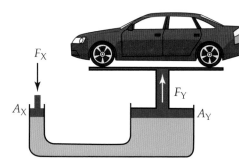

pressure throughout fluid = p_X

▲ **Figure B.3.1.** The hydraulic jack

Figure B.3.1 shows a hydraulic lift used to raise a car. The lift is filled with incompressible oil of constant volume. When a force F_X is applied at X, the pressure p in the liquid is $\dfrac{F_X}{A_X}$. This pressure is transmitted through the fluid so that, at Y, where the area is larger,

$p = \dfrac{F_Y}{A_Y}$. The force exerted upwards on the car is much larger (by $\dfrac{A_Y}{A_X}$) than that exerted at X. However, the volume of fluid moved at X is the same as at Y so the car moves upwards by a small amount (by the ratio $\dfrac{A_X}{A_Y}$). The device is energy neutral assuming no frictional losses.

Buoyancy forces acting upwards on an object submerged in a fluid are described by Archimedes' principle.

Density, $\rho = \dfrac{\text{mass}}{\text{volume}} = \dfrac{m}{V}$

Density is a scalar and has units kg m⁻³.

Pressure

$p = \dfrac{\text{force normal to a surface}}{\text{area of the surface}}$

Pressure is a scalar and has units of N m⁻² or pascal (Pa).

1 N m⁻² = 1 Pa. The fundamental unit of pressure is kg m⁻¹ s⁻².

An important idea in fluid dynamics is **Pascal's principle**. Pascal stated that a pressure change that occurs anywhere in a *confined incompressible* fluid is transmitted through the whole fluid. The key words here are *incompressible* and *confined*.

Example B.3.1

Roughly 90% of the volume of an iceberg is below the water surface. Estimate the density of sea ice.

Density of water = 1000 kg m⁻³

Solution

Consider a 1000 kg slab of ice. By Archimedes' principle, the upthrust on this ice is 900g as this is the weight of water displaced. However, because the volume of the ice is 1000 m³, the density of the ice is 0.9g ≈ 900 kg m⁻³

An ideal, non-viscous incompressible fluid undergoes streamline flow. This means that the motion of a particle at a point in the fluid is the same as the motion of particles that preceded it at the same point. A collection of streamlines constitute a *stream tube*. The stream tube changes dimensions, but the amount of fluid inside it does not as fluid cannot enter or leave the surface of a stream tube. This leads to the *continuity equation*.

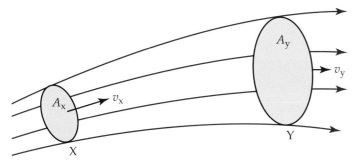

▲ **Figure B.3.2.** Streamlines, stream tube and the continuity equation

The continuity equation describes the changes to the motion of fluid inside a stream tube. As fluid cannot enter or leave, the mass entering and leaving the stream tube must be constant. So, for time Δt, the mass entering and leaving is $Av\rho\Delta t$. The fluid is incompressible, so ρ is constant and the time is the same for both ends of the stream tube.

The Bernoulli equation can be modified by multiplying each term by volume V. So $\frac{1}{2}mv^2 + mgz + pV = \text{constant} \times V$. When this is done, the equation becomes

kinetic energy + gravitational potential energy + work done on fluid = constant

The Pitot–static tube system is used to measure fluid speeds. Two tubes are used: one parallel to a streamline (the static tube) and another with an opening at 90° to the same streamline (the Pitot tube). Assuming that the two tubes are far apart—so that the static tube cannot affect the Pitot tube—when the flow is steady, the pressure difference is a measure of the kinetic energy of the fluid:

$$p_x + \frac{1}{2}\rho v_x^2 = p_y \Rightarrow v_x = \sqrt{2g(h_y - h_x)}.$$

Pitot tubes can be mounted on aircraft wings to measure the aircraft speed relative to the air speed.

Example B.3.2

A tank contains two liquids, A and B. A is vertically above B.

Calculate the speed with which liquid B emerges from a hole at the bottom of the tank.

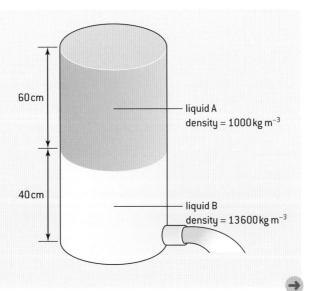

Archimedes' principle states that for an object wholly or partly in a fluid, an upward buoyancy force acts on the object that is equal to the weight of fluid displaced by the object.

The buoyancy force B is usually known as the **upthrust**.

Upthrust is given by $B = \rho_f V_f g$ where ρ_f is the density of the fluid and V_f is the volume of fluid displaced by the object.

An **ideal fluid** has no resistive force either to the fluid itself moving, or to solids moving through it. Such a fluid is **non-viscous** (the meaning of viscosity is considered later).

Continuity equation: $Av = \text{constant}$ and $A_x v_x = A_y v_y$.

Av is known as the volume flow rate and has the units $m^3\,s^{-1}$.

The **Bernoulli equation**,

$$\frac{1}{2}\rho v^2 + \rho gz + p = \text{constant}, \text{ is}$$

a statement of the conservation of energy in the context of fluid dynamics. Each term in the equation has the unit of pressure, $N\,m^{-2}$ or Pa.

This leads to $p = p_0 + \rho_f gd$ for a fluid at rest when comparing pressure at a point in the fluid and the pressure a height d above it.

Solution

The pressure on B due to A is $\rho g h = 1000 \times 9.81 \times 0.6 = 5886\,\text{Pa}$

The pressure at the base of the tank is $13\,600 \times 9.81 \times 0.4 + 5886$
$= 53\,310 + 5880 = 59\,250\,\text{Pa}$. This is equivalent to 0.44 cm of liquid B alone.

Rearranging $\frac{1}{2}\rho v^2 + \rho g z + p = \text{constant}$ and recognizing that p is the same on both sides of the equation gives

$v = \sqrt{2gh} = 2.9\,\text{m s}^{-1}$.

Example B.3.3

Water flows into a long gardening hose of diameter 1.5 cm with a speed of $5.0\,\text{m s}^{-1}$. The spray nozzle has a diameter of 8 mm.

Calculate the velocity of water leaving the nozzle.

Solution
The continuity equation is $Av = \text{constant}$ as density is constant.

So $A_{in}v_{in} = A_{out}v_{out}$

and $v_{out} = A_{in} \times v_{in} A_{out} \left(\dfrac{0.75 \times 10^{-2}}{4.0 \times 10^{-3}} \right)^2 \times 5.0 = 18\ \text{m s}^{-1}$

Viscosity, η is the resistance of a fluid to stress, where one plane of the liquid slides relative to another parallel plane.

The unit of viscosity is the Pa s or N m^{-2} s. In fundamental units, this is kg m^{-1} s^{-1}.

Real fluids have viscosity which varies according to temperature and the chemistry of the fluid. Some fluids are more viscous than others (compare the flow of honey with water at the same temperature, for example).

George Stokes analysed the resistive (drag) force F_D acting on a sphere of radius r due to a viscous fluid while the sphere falls through the fluid under conditions of streamline flow. When an object falls in a resistive medium, it will reach a terminal speed.

Stokes's law: when the speed of the sphere is v, the resistive force
$F_D = 6\pi r v \eta$.

 You saw terminal speed in Topic 2.2.

Including the effects of buoyancy, when the density of the medium is ρ and the density of the sphere is σ, **the terminal speed**
$v_t = \dfrac{2r^2 g (\sigma - \rho)}{9\eta}$.

Example B.3.4

A sphere is falling at terminal speed through a fluid.

The following data are available.

Diameter of sphere	= 3.0 mm
Density of sphere	= 2500 kg m^{-3}
Density of fluid	= 875 kg m^{-3}
Terminal speed of sphere	= 160 mm s^{-1}

Determine the viscosity of the fluid.

Solution
Rearranging the Stokes's law expression gives

$$\eta = \frac{2r^2 g (\sigma - \rho)}{9 v_t} = \frac{2 \times \left(1.5 \times 10^{-3}\right)^2 \times 9.81 \times (2500 - 875)}{9 \times 0.16} = 50\,\text{mPa s}$$

Laminar flow is steady, predictable streamline flow where the stream tubes remain intact and material does not cross between streamlines.

Turbulent flow is unpredictable; it is characterized by the appearance of eddies and vortices in the fluid.

Steady (streamline) flow is observed at low fluid speeds. As the speed increases, the flow between layers of fluid sliding past each other becomes unstable; particles from different streamlines and stream tubes begin to interact and mix. The flow is no longer *laminar* but is now *turbulent*.

To see an example of laminar flow turning into turbulent flow, watch the smoke rising from a piece of smouldering wood or paper. The smoke begins with an orderly flow but a few centimetres above the source of the smoke the flow becomes turbulent.

This transition between the two flow states and the transition speed are difficult to predict. There is a 'rule of thumb' for flow in a pipe using the *Reynolds number R_e*.

When R_e is less than 1000, flow can be taken to be laminar; when R_e is greater than 2000, the flow will be turbulent. These values are different because there is a complex transition between the flow states—it is not possible to be precise about the nature of the flow in the transition.

> The **Reynolds number**, R_e is a dimensionless quantity given by
> $$R_e = \frac{vr\rho}{\eta}$$
> where r is the radius of the pipe, v is the speed of flow, ρ is the density of the fluid, η is its viscosity.

Example B.3.5

A syrup flows at a rate of $4.0\,m^3\,s^{-1}$ in a circular pipe of diameter 6.0 cm. The syrup has a density of $1300\,kg\,m^{-3}$ and a viscosity of 17 Pa s.

Deduce whether the flow is laminar.

Solution

$$\text{Speed of flow} = \frac{\text{flow rate}}{\text{area of pipe}} = \frac{4.0}{\pi \times 0.03^2} = 1.4 \times 10^3 \text{ m s}^{-1}$$

$$R_e = \frac{vr\rho}{\eta} = \frac{\left(1.4 \times 10^3\right) \times 0.03 \times 1300}{17} = 3200$$

This is greater than 2000, so the flow will be turbulent.

SAMPLE STUDENT ANSWER

A ball is moving in still air, spinning clockwise about a horizontal axis through its centre. The diagram shows streamlines around the ball.

a) The surface area of the ball is $2.50 \times 10^{-2}\,m^2$. The speed of air is $28.4\,m\,s^{-1}$ under the ball and $16.6\,m\,s^{-1}$ above the ball. The density of air is $1.20\,kg\,m^{-3}$.

Estimate the magnitude of the force on the ball, ignoring gravity.

[2]

This answer could have achieved 1/2 marks:

$A = 2.5 \times 10^{-2}\,m^2$ Bernoulli equation excluding z because height of the ball is so negligible.

$$P_1 + \frac{1}{2}\,PV_1^2 = P_2 + \frac{1}{2}\,PV_2^2$$

$$P_1 - P_2 = \frac{1}{2} \times 1.2 \times (28.4)^2 - \frac{1}{2} \times 1.2 \times (16.6)^2 = 318.6\,Pa = \Delta p$$

$$P = \frac{F}{A} \qquad F = PA \qquad \Delta F = \Delta P.A$$

$$\Delta F = 318.6 \times 2.5 \times 10^{-2}$$

$$= 7.965 \approx 8\,N$$

▲ The basic method is correct including the decision to ignore the thickness of the ball in the Bernoulli equation (reducing it to two terms).

▼ However, the wrong area is used in the equation and so the answer is incorrect. The cross-sectional area of the ball is $\frac{1}{4}$ of the surface area.

▼ This is a repeat of the comment made in the previous part but is not a good enough answer. The essential point is that, for the Bernoulli equation to hold, the flow must be laminar. When the flow is turbulent, the streamlines will break down and the pressure across the ball will equalize.

← b) State one assumption you made in your estimate in part a).   [1]

This answer could have achieved 0/1 marks:

That the height of the ball is negligible, so the difference in values would be 0.

B.4   FORCED VIBRATIONS AND RESONANCE (AHL)

You must know:

✔ what is meant by the natural frequency of vibration

✔ what is meant by damping, and under-, over- and critically-damped oscillations

✔ what is meant by the Q factor

✔ what is meant by a periodic stimulus and by driving frequency

✔ what meant by resonance in response to a periodic stimulus.

You should be able to:

✔ describe examples of under-, over- and critically-damped oscillations

✔ describe, using a graph, how the amplitude of vibration varies with driving frequency for an object close to its natural frequency of vibration

✔ describe the phase relationship between frequency of a periodic stimulus and the forced oscillations that result from it

✔ solve problems involving Q

✔ describe both useful and destructive effects of resonance.

🔗 Simple harmonic motion is covered in Topics 4.1 and 9.1.

Under-damped oscillators lose energy gradually and come to rest taking many oscillations to do so.

Critically damped oscillators stop moving (and therefore lose their total kinetic energy) in the shortest time possible.

Over-damped oscillators stop moving in a longer time than the critically damped case.

(See Figure B.4.1.)

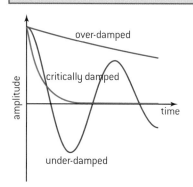

▲ **Figure B.4.1.** *Amplitude* against *time* graph for oscillators with various degrees of damping

Simple harmonic motion has a constant amplitude and energy; this was an implicit assumption of earlier topics. In real cases, a freely oscillating system gradually loses energy through resistance. The frictional losses are known as *damping*.

In under-damping, the amplitude of the oscillation decreases exponentially. This implies that the time to lose half the amplitude (and three-quarters of the energy) is a constant for the system.

A child on a swing is given a push to begin the oscillation. Without intervention, the swing will eventually stop moving. Add energy to the system in a systematic way and the oscillation can be maintained indefinitely. However, when the *driving frequency* (the frequency applied to the swing) does not match the *natural frequency* of the swing, the swing amplitude will be larger or smaller depending on how close the driving frequency is to the natural frequency. The driven system is said to undergo *forced oscillations*.

Figure B.4.2 shows how the amplitude of the driven system varies with frequency of the driver.

An important feature of these curves is that there is a maximum amplitude for each level of damping; this occurs at *resonance*.

Figure B.4.3 shows the phase relationship between the driver system and the driven system. When the driver frequency is greater than driven, then the driver is 180° out of phase with the driven system; when the driver frequency is smaller than the natural frequency, the driver is in phase. At resonance, the driver and the driven system are 90° apart with the driver leading.

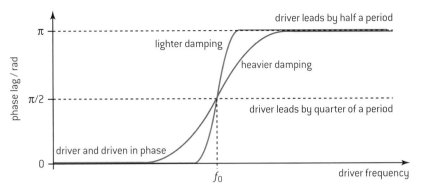

▲ **Figure B.4.3.** Phase relationships between driver frequency and natural frequency

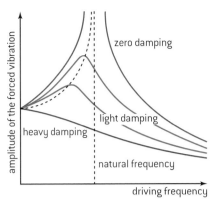

▲ **Figure B.4.2.** Resonance curves for different degrees of damping

Resonance is an important phenomenon. It is used to tune radios, and tides are caused by resonances in the oceans driven by the gravitational attraction of the Sun and Moon. Disadvantages of resonance include large and potentially destructive amplitudes in mechanical systems when an oscillating system is driven close to its natural frequency by a varying driving system.

The width of a resonance curve depends on the amount of damping in the system. This sharpness is defined using a quantity called the Q or "quality" factor. Q is the approximate number of oscillations that a system takes to decay to zero when the driver has been removed.

$$Q = 2\pi \times \frac{\text{energy stored per cycle}}{\text{energy dissipated per cycle}}$$

$$= 2 \times f_0 \times \frac{\text{energy stored}}{\text{power loss}},$$

where f_0 is the resonant frequency.
Q has no unit.

A high Q value implies that a system is lightly damped and will take many cycles to stop moving as the power loss is small. Low Q values indicate that damping is heavy. Typical Q values are 0.5 for the damping system on a door closure, 100 for a simple pendulum in air, and 30 000 for the vibrating crystal in a quartz watch.

Example B4.1

An electric motor maintains the oscillation of a demonstration simple pendulum that has a time period of 10.0 s. The power output of the motor is 48 mW and the maximum kinetic energy of the motion of the pendulum bob is 2.3 J.

The motor is switched off. Deduce the time the system will take to stop.

Solution
The pendulum is storing energy at the rate of 2.3 W with power being supplied at the rate of 48 mW.

The resonant frequency is 0.1 Hz.

So $Q = 2\pi \times 0.1 \times \dfrac{2.3}{4.8 \times 10^{-2}} = 30$

The system is going to oscillate about 30 times before stopping; in other words, in about 5 minutes.

▲ (The upper curve was provided.) The lower curve is reasonably drawn and incorporates the features that examiners were looking for. The peak is lower than the original because the damping is greater; also the peak is shifted to lower frequencies, which is also correct.

▼ The endpoints are not clear; on the left, the curves appear to cross. Try to avoid this ambiguity. There was no penalty, however.

A driven system is lightly damped. The graph shows the variation with driving frequency f of the amplitude A of oscillation.

a) On the graph, sketch a curve to show the variation with driving frequency of the amplitude when the damping of the system increases. [2]

This answer could have achieved 2/2marks:

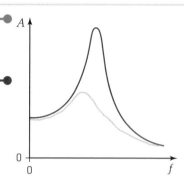

b) A mass on a spring is forced to osciiate by connecting it to a sine wave vibrator. The graph shows the variation with time t of the resulting displacement y of the mass. The sine wave vibrator has the same frequency as the natural frequency of the spring–mass system. [2]

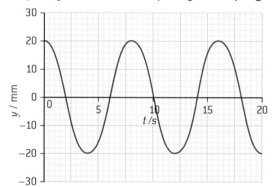

i) State and explain the displacement of the sine wave vibrator at $t = 8.0$ s. [2]

This answer could have achieved 2/2 marks:

▲ The answer begins by correctly quoting the phase relationship at resonance. It is important to use the word resonance in the answer because it is not mentioned in the question and the condition of resonance has to be identified. The answer goes on to make correct deductions: the spring is at a maximum at 8.0s; therefore, the vibrator must be at zero displacement.

> Because it is resonating, the phase difference will be $\frac{\pi}{2}$.
>
> Therefore, the sine wave vibrator at 0.80 seconds will be at one of its zeroes, which means the displacement s = 0.
>
> Displacement = 0.

ii) The vibrator is switched off and the spring continues to oscillate. The Q factor is 25.

Calculate the ratio $\dfrac{\text{energy stored}}{\text{power loss}}$ for the oscillations of the spring–mass system. [2]

This answer could have achieved 2/2 marks:

▲ The answer uses all the data provided including the frequency from earlier to provide a correct solution to the ratio.

> $Q = 2\pi.f_a \dfrac{E\ stored}{Powerless}$ $T = 8\ seconds$ $f = \dfrac{1}{T} = 0.125\,Hz = f_a$
>
> $25 = 2\pi \times 0.125 \times r$ $r = 31.8$ $\dfrac{Energy\ stored}{Power\ loss} = 31.8$

Practice problems for Option B

Problem 1

A potter's wheel is rotating at an angular speed of $5.0 \, \text{rad s}^{-1}$ with no torque acting.

The potter throws a lump of clay onto the wheel so that the wheel and the clay have a common axis of rotation.

The moment of inertia of the wheel is $1.6 \, \text{kg m}^2$ and the moment of inertia of the clay is $0.25 \, \text{kg m}^2$, both about the common axis.

The angular speed of the wheel changes suddenly when the clay lands on it and no net angular impulse is added to the system.

a) Calculate the angular speed of the wheel immediately after the clay has been added.

The potter now applies a tangential force to the rim of the wheel over 0.25 of a revolution to return the angular speed to $5.0 \, \text{rad s}^{-1}$.

The wheel has a diameter of 0.62 m.

b) Calculate the angular acceleration of the wheel.

c) Deduce the average tangential force applied by the potter.

Problem 2

Explain why a piece of wood floats.

Problem 3

A Pitot–static tube is used in an aircraft that is travelling at $75 \, \text{m s}^{-1}$ into a headwind of $15 \, \text{m s}^{-1}$. The density of the air is $1.3 \, \text{kg m}^{-3}$.

Determine the pressure difference measured by the instrument.

Problem 4

a) A mass of 24 kg is attached to the end of a spring of spring constant $60 \, \text{N m}^{-1}$. The mass is displaced 0.035 m vertically from its equilibrium position and released.

Determine the maximum kinetic energy of the mass.

b) The mass–spring system is damped and its amplitude halves by the end of each complete cycle.

Sketch a graph to show how the kinetic energy of the mass on the spring varies with time over a single period. You should include suitable values on each of your scales.

Problem 5

A flywheel consists of a solid cylinder of mass 1.22 kg and radius 240 mm.

A mass M is connected to the flywheel by a string wrapped around the circumference of the cylinder.

The mass falls from rest and exerts a torque on the flywheel which accelerates uniformly.

After a time of 4.85 s, the velocity of M is $2.36 \, \text{m s}^{-12}$.

Moment of inertia for the flywheel $= \dfrac{1}{2} \, mass \times radius^2$

a) Calculate the angular acceleration of the flywheel.

b) Deduce the torque acting on the flywheel.

c) Determine M.

Problem 6

An ideal gas is compressed quickly by a piston in a cylinder. The gas is initially at a pressure of 110 kPa and a temperature of 290 K. The volume of the gas is $6.0 \times 10^{-4} \, \text{m}^3$.

a) Calculate the quantity of gas in the cylinder. State an appropriate unit for your answer.

b) The gas is compressed to a pressure of 190 kPa and a volume of $4.0 \times 10^{-4} \, \text{m}^3$

(i) Suggest, with a calculation, whether the gas is compressed isothermally.

(ii) Explain why the compression may be adiabatic.

c) 15 J of energy is used to compress the gas.

Determine the change in the internal energy of the air in the cylinder.

d) The compression is repeated very slowly.

Discuss the entropy change that takes place in the cylinder and its surroundings as the air is compressed.

C IMAGING

You must know:

✔ the definition of a real image and virtual image

✔ the definition of linear magnification and virtual magnification

✔ the optical properties of converging and diverging mirrors

✔ the definition of focal point, focal length and radius of curvature for curved mirrors

✔ the optical properties of converging and diverging thin lenses

✔ the definition of focal point, focal length and principal axis for converging and diverging lenses.

You should be able to:

✔ construct scaled ray diagrams to solve problems for converging and diverging mirrors

✔ describe how the properties of a thin lens arise from the way the curved surfaces modify wave fronts incident on them

✔ identify the principal axis focal point and focal length of a converging or diverging lens on a scaled diagram

✔ construct scaled ray diagrams to solve problems for converging and diverging lenses

✔ solve problems involving the thin lens equation and linear and angular magnification

✔ explain spherical aberration and chromatic aberration and describe how the ir effects can be reduced.

Reflecting mirrors with spherical surfaces modify wave fronts of light incident on them to form images. These surfaces can be either concave (a converging mirror) or convex (a diverging mirror). Figure C.1.1 shows the features of both types of mirror.

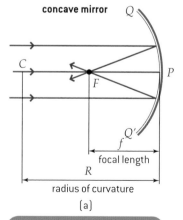

(a)

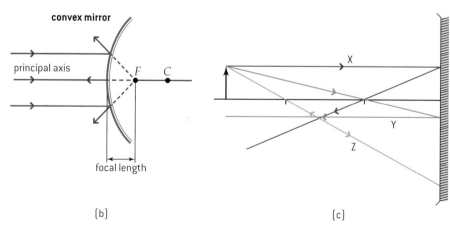

(b)   (c)

▲ **Figure C.1.1.** Concave and convex mirrors

>> **Assessment tip**

You must know the meanings of **principal axis** PC, focal point F, **focal length** f, **centre of curvature** C and **radius of curvature** R, as defined on the diagrams.

The focal length is half the radius of curvature for both mirrors.

Figure C.1.1(a) shows the path of three rays incident on the converging mirror that are parallel to the principal axis (one ray coincident with it). These rays are said to come from infinity and they *converge* (come to a *focus*) at the focal point.

Figure C.1.1(b) for a diverging mirror also shows three rays, but this time they move apart after reflection—they *diverge*, appearing to have come from the focal point.

Ray diagrams are used to determine the image position formed by a mirror. They use rays whose behaviour is known.

Figure C.1.1(c) shows three such rays for a converging mirror. The three rays are *predictable*; in other words, they follow set rules.

> On a **ray diagram**:
> - ray X travels first parallel to the principal axis and after reflection goes through F
> - ray Y arrives through F and then reflects parallel to the principal axis
> - ray Z goes through C and then returns along its original path.
>
> Arrows are conventionally drawn on the rays to show the direction of travel.
>
> **Rays from infinity** are imagined as originating from the same point on the object and travelling such a long distance that they become parallel to each other.

Figure C.1.2(a) shows rays for a converging mirror. The image formed is smaller than the object (diminished) and is real.

Figure C.1.2(b) shows the ray diagram for a diverging lens. A diminished virtual image is always produced with this type of mirror.

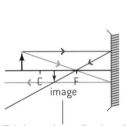

This image is smaller than the object (diminished), upside down (inverted) and real (because the rays cross).

(a)

This image is smaller than the object (diminished), the right way up (erect) and virtual (because the rays appear to have come from the image and do not cross).

(b)

▲ **Figure C.1.2.** Ray diagrams for (a) converging (b) diverging mirrors

> **Magnification** M is a property of both mirror and lens systems. It is the ratio of image size to object size:
>
> $$M = \frac{\text{height of image}}{\text{height of object}}.$$
>
> where v and u are the distances from the mirror to the image and object, respectively, and θ_i and θ_o are the angles subtended at the mirror by the rays from the top of the image and object, respectively.
>
> Magnification has no units.
>
> For simple systems
>
> $$M = \frac{\text{image distance}}{\text{object distance}} = \frac{v}{u} = \frac{\theta_i}{\theta_o}$$

Spherical mirrors distort images. These distortions are known as *spherical aberrations*. Aberration can be eliminated using a parabolic mirror where all rays parallel to the principal axis are reflected to the focus.

> **≫ Assessment tip**
>
> It is best not to draw the mirrors as curved surfaces—they are effectively flat compared with the distances between the mirror, image and object. Figure C.1.2 shows the symbols often used for the two types of mirror.
>
> A similar convention is used for lenses (Figure C.1.3).

> **Real images** can be formed on a screen. They form where rays meet and cross.
>
> **Virtual images** cannot be formed on a screen. They are formed by rays that appear to have come from a point. A focusing system is needed to view them.

> **≫ Assessment tip**
>
> Magnification does not always mean that the image is larger than the object—it is the *ratio* of image size : object size. So it can be less than one. In this case, the term *diminished* can be used to describe the image.

Example C.1.1

Construct the ray diagram for a converging mirror where the object is:

a) between F and C   c) at F

b) at C   d) between the pole of the mirror and F.

Solution

(a)

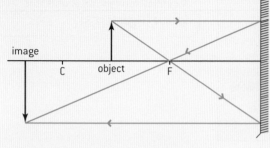

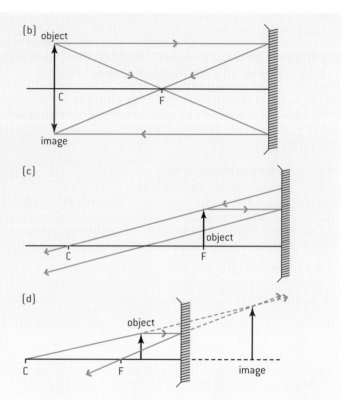

> The physics of refraction at a single plane interface between two media is covered in Topic 4.4. Effects here are extensions to earlier work as there are two interfaces to the lens and both are curved.

Converging and diverging lenses have a technical language similar to mirrors. Figure C.1.3 shows the important quantities associated with the lenses.

The lenses refract the light. As a plane wavefront moves into the lens, the wavefront:

• slows down in the lens

• must be continuous at the interface.

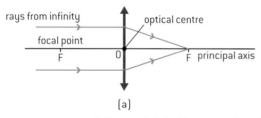 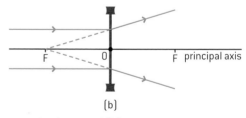

▲ **Figure C.1.3.** The terms for a (a) converging lens and (b) a diverging lens

This means that the wavefronts curve in the lens and converge to a point which is the focus (Figure C.1.4). The effect is reinforced when the wavefront exits the second surface of the lens.

The assumption used in the DP physics course is that the lenses are thin. This is emphasised by the symbols used for the two types of lens (Figure C.1.3). When the lens has significant thickness, different equations must be used to allow for this. Predictable rays exist for the lenses.

▲ **Figure C.1.4.** Wavefronts modified by a single curved interface

Ray ① parallel to the principal axis is refracted through the focal point (diverging lens: moving away from the focal point).

Ray ② through the optical centre of the lens is undeviated.

Ray ③ through the focal point travels parallel to the principal axis after refraction (for a diverging lens, a ray aimed at the focal point is refracted parallel to the principal axis).

Figure C.1.5 shows the ray diagrams for both converging and diverging lenses.

>> **Assessment tip**

Make sure that you can draw all these diagrams confidently and accurately. The same advice applies to the more complicated diagrams in Option C.2. There is advice there on how to remember the constructions but you might consider a similar approach to your learning here.

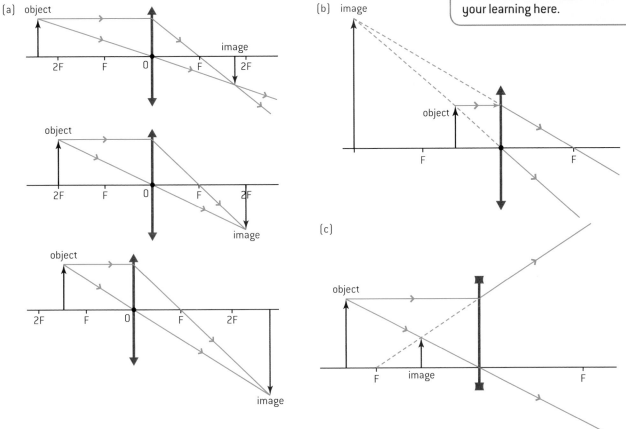

Figure C.1.5. Ray diagrams (a) and (b) for a converging lens, and (c) a diverging lens

There are also numerical methods to predict the behaviour of a lens system using the *thin-lens equation*.

The thin-lens equation links to the ideas of wavefront modification by the lens. The more strongly curved the surfaces, the greater the change to the wavefront shape, and the smaller the value of f for the lens. Therefore, a small radius of curvature means a small f. This leads to the *power* of a lens to alter waves; a powerful lens has a small f.

The thin-lens equation requires the use of a sign convention to distinguish mathematically between real and virtual objects and images and between the focal lengths of converging and diverging lenses.

The **thin-lens equation** is

$$\frac{1}{f} = \frac{1}{u} + \frac{1}{v}$$

where u is the distance of the object (from the optical centre of the lens), v the image distance and f the focal length.

Example C.1.2

A lens of power +4D is placed 75 cm from an object of height 5.0 cm.

Calculate:

a) the position of the image formed

b) the nature of the image.

>> **Assessment tip**

The thin-lens equation involves reciprocals and a sign convention. Take care with both, particularly with the reciprocal at the end of the calculation.

The **power** P of a lens is defined as

$$P \propto \frac{1}{\text{radius of curvature of surfaces}}$$
$$= \frac{1}{f}$$

The power is measured in dioptres (D) and has the unit m^{-1}. A lens with $f = 0.25$ m has a power +4D.

The human eye has a **near point**. It is the closest distance D at which we can see an object without strain. The near point is taken to be 25 cm for the normal eye.

Solution

a) The focal length f is $\frac{1}{4}$ and the lens is converging. $f = +0.25$ m.

   The object distance is 0.75 m.

   Rearranging $\frac{1}{f} = \frac{1}{u} + \frac{1}{v}$ gives $\frac{1}{v} = \frac{1}{f} - \frac{1}{v} = \frac{1}{+0.25} - \frac{1}{0.75}$

   $$= 2.6\dot{6} = \frac{1}{0.375}$$

   So the image distance is +0.375 m = 37.5 cm from the lens.

b) The answer to part a) was positive, so the image is real.

   Its magnification is $\frac{0.375}{0.75} = \frac{1}{2}$, so the image height is $5 \times \frac{1}{2} = 2.5$ cm.

Figure C.1.5 (b) shows how a converging lens produces a magnified virtual image. This is the magnifying glass, one of the oldest optical instruments.

A magnifying glass can be used:

- with the object at f and the image at infinity; this gives an angular magnification $m = \frac{\theta_i}{\theta_o}$ where θ_i and θ_o are the angles subtended at the eye by the image and object respectively, leading to: $M_{\text{infinity}} = \frac{D}{f}$
- with the object closer than f and the image at the near point; this gives an angular magnification: $M_{\text{near point}} = \frac{D}{f} + 1$.

Example C.1.3

A simple magnifying glass has a focal length of 0.10 m. It is used by a person with a near point of 0.30 m. Calculate the angular magnification of the object when the image is formed at the person's near point.

Solution

$$M = \frac{D}{f} + 1 = \frac{0.30}{0.10} + 1 = 4\times$$

The angle from top to bottom of the image is four times greater than that for the object.

Lenses, like mirrors, are prone to aberration. Unlike mirrors, lenses have two forms of aberration: spherical (due to the physical shape of the lens) and chromatic (due to the variation of refractive index with wavelength in the lens). Figure C.1.6 shows how the aberrations form.

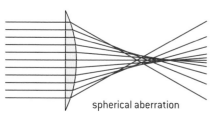

spherical aberration

▲ Figure C.1.6. Aberrations produced by a lens

Chromatic aberrations are caused by the variation in wave speed with wavelength (colour) of the light. Red light travels faster than blue light in the lens, so the red and blue wavefronts emerge with different shapes and give slightly different focal lengths. This leads to colour fringing at the image because the magnifications are different.

The aberration can be reduced by using an achromatic doublet where two lenses—a converging lens and a diverging lens—are combined. They are made from different refractive indices and so produce the required power while the aberrations are in opposite directions and cancel out.

SAMPLE STUDENT ANSWER

A lamp is located 6.0 m from a screen.

Somewhere between the lamp and the screen, a lens is placed so that it produces a real inverted image on the screen. The image produced is 4.0 times larger than the lamp.

6.0 m

lamp

a) Determine the distance between the lamp and the lens. [3]

This answer could have achieved 2/3 marks:

$$M = \frac{-v}{u} \qquad v + u = 6m$$

$$-4 = \frac{-v}{u}$$

$$4v = u \rightarrow 5v = 6m$$

$$v = 1.2m$$

$$u = 6 - 1.2 = 4.8m$$

▼ The answer begins with an unfortunate slip. When $4 = \frac{v}{u}$ then $4u = v$, not the other way round.

▲ However the rest of the solution is correct. The answer shows that the total distance between the lamp and the screen is $5v$ (following the error which the examiner will carry forward) and hence u, in this case, must be 4.8 m.

b) Calculate the focal length of the lens. [1]

This answer could have achieved 1/1 marks:

$$\frac{1}{f} = \frac{1}{u} + \frac{1}{v}$$

$$4.8 \quad 1.2 \qquad f = 0.96m$$

▲ Fortunately, the use of the thin-lens equation is not affected by the error in the previous question and scores full marks.

c) The lens is moved to a second position where the image on the screen is again focused. The lamp–screen distance does not change. Compare the characteristics of this new image with the original image. [2]

This answer could have achieved 1/2 marks:

The new image is still real and inverted.

▼ There is detail missing here. The facts needed are: Is the image real or virtual? Is it upright or inverted? Is it magnified or diminished? In this case, the last point is omitted—the answer is that the magnification is $\frac{1}{4}$.

C.2   IMAGING INSTRUMENTATION

You must know:

✔ how optical compound telescopes form images

✔ how astronomical refracting telescopes form images

✔ how astronomical reflecting telescopes form images

✔ what are meant by the Newtonian mounting and the Cassegrain mounting for reflecting telescopes

✔ the operation and resolution of a single dish radio telescope

✔ the definition of a radio-interferometer telescope

✔ how to describe the comparative performance of Earth-based and satellite-based telescopes.

You should be able to:

✔ construct and interpret the ray diagrams for a compound microscope in normal adjustment

✔ construct or complete the ray diagram for an astronomical refracting telescope in normal adjustment

✔ investigate the performance of an optical compound microscope and an astronomical refracting telescope

✔ solve problems involving the angular magnification and resolution of an optical compound microscope

✔ solve problems involving the angular magnification of optical astronomical telescopes.

> 🔗 The meaning of the term *near point* is discussed in Option C.1.

The magnifying glass described in Option C.1 has disadvantages. Small focal lengths are required for large magnifications and the large curvatures required means that spherical aberrations are a problem. The optical compound microscope solves this; *compound* means that there is more than one lens—a converging *objective* lens and a converging *eyepiece* lens. You only need to know the *normal adjustment* when the image is formed at the near point of the observer's eye.

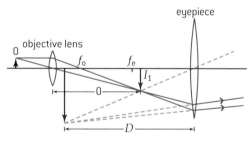

▲ **Figure C.2.1.** Compound microscope in normal adjustment

Figure C.2.1 shows the ray diagram for a compound microscope in normal adjustment.

- The **object** is to the left of the **objective lens** which forms a real magnified image I_1 of the object at a point between the focal point of the eyepiece and the eyepiece itself.

- The **eyepiece** acts as a magnifying glass viewing I_1 to produce a virtual, magnified image of the intermediate I_1.

- The distance between the virtual image and the eyepiece is D; the observer's eye should be placed as close to the eyepiece as possible.

The angular magnification of the compound microscope is given by multiplying the magnifications of the objective and eyepiece lenses.

This is $\left(\dfrac{v_o}{u_o}\right)\left(\dfrac{D}{f_e}+1\right) \approx \dfrac{DL}{f_o f_e}$

where L is the length of the microscope tube, defined using

$$\text{overall magnification} = \dfrac{\text{linear magnification}}{\text{of eyepiece}} \times \dfrac{\text{angular magnification}}{\text{of objective}}$$

and $\dfrac{v_o}{u_o} \approx \dfrac{v_o}{f_o}$

The microscope just resolves two images when (following the Rayleigh criterion) $\sin\theta = 1.22\dfrac{\lambda}{d}$, where θ is the angle subtended at the eye by the images and d is the effective diameter of the aperture (this is usually the objective). Resolution is improved by using short wavelengths of light and as wide an aperture as possible.

Example C.2.1

A compound microscope is in normal adjustment with the objective and eyepiece lenses separated by 23 cm. The object is 6.2 mm from the objective which has a focal length of 6.0 mm. The focal length of the eyepiece is 50 mm.

Determine the magnification of the microscope.

Solution
An intermediate image is formed in the microscope tube at distance x from the objective:

$$\frac{1}{6.2} + \frac{1}{x} = \frac{1}{6}$$

So, $x = \left(\dfrac{1}{6} - \dfrac{1}{6.2}\right)^{-1} = 186$ mm from the objective.

The linear magnification of the objective is $\dfrac{186}{6.2} = 30\times$

And the angular magnification of the eyepiece is $\left(\dfrac{250}{50} + 1\right) = 6\times$

So the overall magnification is $30 \times 6 = 180\times$

>> **Assessment tip**

Learn the function of the individual elements in the optical instruments. For example, a compound microscope can be regarded as a modification of the magnifying glass (the eyepiece) with the objective lens providing a two-stage magnification process that avoids the magnifying glass limitations.

To draw the instruments in an examination, begin with the final image and work backwards. Add the intermediate images at this point, and then the initial rays. Make sure that you practice drawing them.

The astronomical refracting telescope also consists of two converging lenses. The focal length of the objective is longer than the focal length of the eyepiece and, in normal adjustment, the focal points coincide. The length of the tube is $f_o + f_e$. Figure C.2.2 shows the ray diagram for an astronomical telescope in normal adjustment.

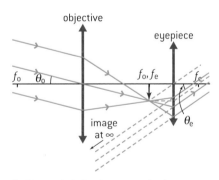

▲ **Figure C.2.2.** Astronomical telescope in normal adjustment

- The object is at infinity, so the rays incident on the objective lens are parallel.

- These rays are focused by the objective lens at its focal length f_o to form a real image of the object at the focus.

- This real image is a real object for the eyepiece lens of focal length f_e. This lens forms a final image at infinity from this object.

- The two lenses magnify the object separately. The overall angular magnification and magnification of the telescope is $M = \dfrac{\theta_o}{\theta_e} = \dfrac{f_o}{f_e}$.

Example C.2.2

An astronomical telescope in normal adjustment has two lenses of focal lengths +96 cm and +12 cm.

a) Calculate:

  i) the length of the telescope

  ii) the angular magnification of the telescope.

The telescope is used to view the Moon which subtends an angle of 0.52° at the Earth.

b) Determine the diameter of the image of the Moon as formed by the telescope.

Solution

a) i) The total length $= f_o + f_e = 108$ cm

  ii) Angular magnification $= \dfrac{f_o}{f_e} = \dfrac{96}{12} = 8\times$

b) The diameter of the image is:

$f_o \times \tan(\text{angle subtended by object}) = 96 \times \tan(0.52) = 0.871$ cm

$$= 8.7\,\text{mm}$$

Another type of astronomical telescope uses reflecting mirrors rather than refracting lenses. Two forms of the astronomical reflecting telescope are required for the DP physics course: the Newtonian mounting (Figure C.2.3a)) and the Cassegrain mounting (Figure C.2.3b)).

(a)
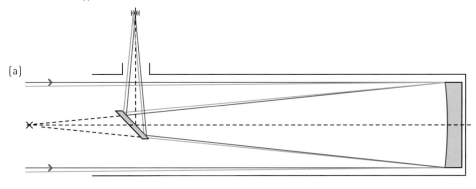

(b)
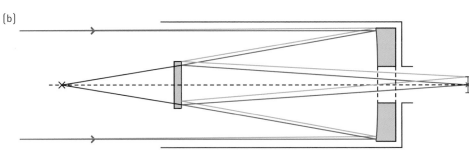

▲ **Figure C.2.3.** Astronomical reflecting telescope (a) Newtonian mounting and (b) Cassegrain mounting

Rays from a distant object are focused by a primary converging mirror to a focus on the principal axis. When a screen or photographic plate is placed at the focus, an image can be seen.

Newtonian mount: The image is viewed through an eyepiece from the side, the rays having been turned by 90° from the principal axis by a plane mirror. This flat surface does not modify the magnification.

Cassegrain mount: A curved hyperbolic mirror sends the rays through a small hole in the primary mirror. The secondary mirror effectively lengthens the main tube and increases the magnification. Again, the final image is viewed using a converging eyepiece lens.

These telescopes do not suffer from chromatic aberration (when no further lens-based imaging is used). Only one main curved surface has to be prepared. However, the surfaces are vulnerable and easily damaged.

Example C.2.3

A Cassegrain telescope primary mirror has a diameter of 0.80 m.

a) Compare the effectiveness of the telescope at collecting energy with a human eye with a pupil diameter of 8.0 cm.

b) The effective focal length of the telescope main mirror is 2.8 m and the focal length of the eyepiece is 7.0 cm.

   Calculate the angular magnification of the telescope.

Solution

a) The area of the mirror is $(10)^2 = 100$ times the area of the eye, so the telescope collects $100 \times$ as much energy every second.

b) Angular magnification $= \dfrac{f_o}{f_e} = \dfrac{280}{7.0} = 40 \times$

Radio telescopes are also reflecting telescopes; they use radio wavelengths for the radiation rather than light.

Single-dish reflectors collect electromagnetic radiation from distant objects and reflect it to the focus of a parabolic dish. All rays parallel to the principal axis are focused at the focus. The larger the dish, the greater the energy collected per second.	**Interferometer telescopes** are a recent development. Signals from a number of small single-dish telescopes (geographically separated) are combined. The system emulates a large single dish of baseline B.
The resolution is given by $\sin\theta = 1.22\dfrac{\lambda}{d}$, so the larger d the better the resolved images.	The resolution is given approximately by $\sin\theta \approx \dfrac{\lambda}{d}$.
Disadvantages of a very large dish are the problems of retaining the parabolic shape in a large structure and the problems of steering a massive object.	The exact resolution depends on the arrangement of small dishes. There are proposals for very large baseline radio telescopes that combine the signals from arrays of dish telescopes in different continents.

Telescopes are now routinely placed on satellites.

• Stars emit radiation some of which is absorbed by the atmosphere. This is not a problem in space.

• The atmosphere distorts telescope images due to refractive index changes and turbulence.

• Extra-terrestrial telescopes are immune to light pollution from cities.

• International collaboration is a feature of satellite development.

Example C.2.4

Two stars that are 2×10^6 light-years apart, emitting light of wavelength 630 nm, are imaged by the Hubble Space telescope which has a primary mirror of 2.40 m diameter.

The light-year is a measure of distance.

a) Determine the angle between two images just resolved by the instrument.

b) Two stars in the Andromeda galaxy are both about three million light years from Earth.

Deduce, in light-years, the separation of these stars so that they can still be resolved.

Solution

a) Using $\sin\theta = 1.22\dfrac{\lambda}{d}$ leads to

$$\theta = \sin^{-1}\left(1.22 \times \frac{6.3 \times 10^{-7}}{2.4}\right) = 3.2 \times 10^{-7} \text{ rad}$$

b) The distance must be $3.2 \times 10^{-7} \times 3 \times 10^{6} = 1$ light-year

---

SAMPLE STUDENT ANSWER

Both optical refracting telescopes and compound microscopes consist of two converging lenses.

a) Compare the focal lengths needed for the objective lens in a refracting telescope and in a compound microscope. [1]

This answer could have achieved 1/1 marks:

> Focal lengths of the objective lens are larger than its eyepiece lens for telescopes whilst focal length of objective lens is smaller than eyepiece lens focal length for microscopes.

▲ The student receives a mark for showing understanding of the focal length ratios for both instruments.

b) A student has four converging lenses of focal length 5, 20, 150 and 500 mm. Determine the maximum magnification that can be obtained with a refracting telescope using two of the lenses. [1]

This answer could have achieved 1/1 marks:

> $M = \dfrac{f_o}{f_e} = \dfrac{500\,mm}{5\,mm} = 100$

» Assessment tip

The question is "determine". It would have been good to see a statement such as "the maximum value for magnification is obtained when the ratio $\dfrac{f_o}{f_e}$ is greatest".

c) There are optical telescopes that have diameters about 10 m. There are radio telescopes with single dishes of diameters at least 10 times greater.

i) Discuss why, for the same number of incident photos per unit area, radio telescopes need to be much larger than optical telescopes. [1]

This answer could have achieved 0/1 marks:

> Because radio telescopes have radio waves of wavelengths that are long and would result in poor resolution. Thus to compensate for this, radio telescopes have to be larger in diameter.

▼ Long wavelengths are identified as leading to poor resolution. The real answer is that the long-wavelength photons each have a small energy. A larger area is needed to collect the equivalent radiation power.

ii) Outline how is it possible for radio telescopes to achieve diameters of the order of a thousand kilometres. [1]

This answer could have achieved 0/1 marks:

> They could create the dish of large diameter in a huge earth crater and by combining many small dishes together.

▼ The question needs an answer that refers to the use of multiple single-dish radio telescopes linked together into an interferometer arrangement. The phrase 'combining many small dishes together' is not clear enough for a mark.

C.3 FIBRE OPTICS

You must know:

✔ the structure of an optic fibre

✔ the definition of a step-index fibre and a graded-index fibre

✔ the definition of waveguide (modal) dispersion and of material dispersion

✔ the meaning of attenuation

✔ the definition of the decibel (dB) scale.

You should be able to:

✔ explain the action of an optical fibre in terms of total internal reflection and critical angle

✔ solve problems in the context of fibre optics using total internal reflection and critical angle

✔ solve problems involving attenuation

✔ describe the advantages of fibre optics over twisted-pair and coaxial cables.

An optical fibre consists of a very thin core of transparent material. It is surrounded by cladding with a lower refractive index than the core. The cladding is usually covered with a protective sheath. When electromagnetic radiation is shone along the core (both light and infra-red radiation are commonly used), total internal reflection occurs at the core–cladding interface so that none of the radiation is lost from the core. The glass has very low attenuation at the wavelength used.

Early optical fibres were *step-index* fibres. However, a *graded-index* fibre, with a gradual reduction in refractive index from the centre to the outside of the core, has advantages for the transmission process. This reduces the effects of *waveguide (modal) dispersion*.

> 🔗 The basic physics behind optical fibres was covered in Topic 4.4.

> 🔗 For total internal reflection,
> the equation $n = \dfrac{1}{\sin c}$, where c
> is the critical angle, follows from
> $\dfrac{n_1}{n_2} = \dfrac{\sin\theta_2}{\sin\theta_1}$, which you met in
> Topic 4.4.

> **Step-index fibres** have a constant refractive index n in the core with an abrupt change of n between the core and cladding.
>
> Figure C.3.1(a) shows the variation of n and the passage of two light rays through a straight fibre. One ray is along the axis the other ray travels a much longer distance than the first ray. A single pulse of light will be broadened in time when it reaches the end (Figure C.3.2). This is **waveguide (modal) dispersion**.
>
> **Graded-index fibres** can correct waveguide dispersion. The refractive index is not constant in the core but varies with distance from the centre, as shown in Figure C.3.1(b). The centre of the core has a larger refractive index compared with the outer part of the core. Large-angle rays now travel faster when in the outer region of the core, whereas rays in the centre travel more slowly.
>
> Smaller core diameters and graded-index materials reduce waveguide dispersion significantly.

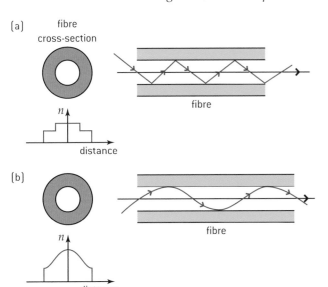

(a) fibre cross-section

fibre

(b)

fibre

▲ Figure C.3.1. (a) Step-index and (b) graded-index optical fibres

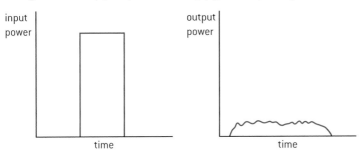

input power

output power

time

time

▲ Figure C.3.2. Changes in the profile of a pulse as it passes along an optical fibre

Assessment tip

The cause of material dispersion links to chromatic aberration and the basic ideas of refraction. Use these common ideas to help build your understanding of these topics.

Transmitted signals can also be distorted through *material dispersion*.

Dispersion broadens the pulse and, when one pulse overlaps the next, electronic systems cannot separate the two pieces of information.

Material dispersion arises because the refractive index of the core depends on wavelength. Suppose the digital signal is transmitted using a ray of white light. The refractive index for red light is less than that for blue, and so red light travels faster in the glass. This means that, the red light will exit first and could 'catch up' with blue light that entered the core earlier.

Material dispersion leads to a spread in the pulse width. The problem can be reduced by restricting the wavelengths used in the core. This restricts the bandwidth and, ultimately, the number of channels available in the fibre.

The signal needs amplification at regular intervals along the fibre to overcome the effects of *attenuation*. The change in intensity is expressed as a ratio using the *Bel scale*.

The **Bel scale** is defined as the logarithm to base 10 ($\log_{10}$) of the ratio of the intensity (or power) of a signal to a reference level of the signal.

$$\text{Attenuation in bel} = \log_{10}\left(\frac{I}{I_0}\right)$$

An attenuation of 5 bel is a power ratio of 10^5, which is large, so the decibel is frequently used.

Attenuation in decibel (dB)

$$= 10\log_{10}\left(\frac{I}{I_0}\right)$$

10 dB is a ten-fold change in intensity. A ratio of two in intensity is about 3 dB.

Because *intensity* is proportional to *amplitude*^{2}, attenuation can also be

$$\text{written as: } 20\log_{10}\left(\frac{A}{A_0}\right)$$

There are two principal causes of attenuation in optical fibres: absorption and scattering.

Example C.3.1

An optical fibre has an attenuation loss of 2.6 dB km^{-1}. The signal needs to be amplified when the power in the signal has been attenuated to 6.0×10^{-15} W.

The input power to the optical fibre is 25 mW.

Deduce the maximum distance between amplifiers for this optical fibre.

Solution
The power ratio is $10\log_{10}\left(\dfrac{2.5 \times 10^{-3}}{6.0 \times 10^{-15}}\right) = 116$ dB.

So, as each kilometre of fibre loses 2.6 dB, an amplifier is needed

every $\dfrac{116}{2.6} = 44.7 \equiv 45$ km.

Optical fibres are now the norm for communication channels involving physical links. They have significant advantages over twisted-pair and coaxial cables.

A twisted-pair arrangement gives only moderate immunity from noise. An external electrical signal will generate similar emfs in both wires and will, therefore, be cancelled out.

Coaxial cables give good immunity to electrical noise. They are used to carry weak signals. However, the cable is bulky and expensive, unlike modern optical fibres.

SAMPLE STUDENT ANSWER

Optical fibres can be classified, based on the way the light travels through them, as single-mode or multimode fibres. Multimode fibres can be classified as step-index or graded-index fibres.

a) State the main physical differences between step-index and graded-index fibres. [1]

This answer could have achieved 1/1 marks:

Step-index fibres have a refractive index for the core and a refractive index for the cladding. Graded-index fibres have varying refractive indexes that decrease outwards from the core to the cladding.

▲ The answer has a good description of the graded-index fibre variation of n.

▼ However, the first sentence – describing the step-index fibre – is poorly expressed. The answer means that the refractive indices for the core and cladding are different and could have added that n_{core} is greater than $n_{cladding}$.

b) Explain why graded-index fibres help reduce waveguide dispersion. [2]

This answer could have achieved 2/2 marks:

Light which travels long pathways are made to travel at faster speeds since graded index fibres have refractive indices that decrease outwards from the core and as refractive index is inversely proportional to speed of light, it causes light which travel extra distances to travel at higher speeds and thus have the light arrive with the same arrival times, to reduce smearing of pulses and reduce waveguide dispersion.

▲ A complete answer that deals with the refractive index variation correctly and goes on to explain how the axial rays (shorter distance) travel more slowly than the high-angle rays (longer physical distance).

C.4   MEDICAL IMAGING (AHL)

You must know:

✔ how medical X-rays are detected and how the X-ray images are recorded

✔ techniques for improving sharpness and contrast in the X-ray image

✔ how ultrasound is generated and subsequently detected in medical contexts

✔ how nuclear magnetic resonance is used to image inside the body

✔ how to explain the use of a gradient field in NMR

✔ the advantages, disadvantages and risks of ultrasound and nuclear magnetic scanning methods and be able to discuss them including a simple assessment of risk.

You should be able to:

✔ explain and solve problems in the context of X-ray imaging involving attenuation coefficient, half-value thickness, linear and mass absorption coefficients

✔ explain, in the context of medical ultrasound, the choice of frequency, the use of gel and the difference between A and B scans

✔ solve problems, in the context of medical ultrasound, involving acoustic impedance, speed of ultrasound through tissue and air, and relative intensity levels of ultrasound

✔ explain the origin of the relaxation of proton spin and how this leads to an emitted signal in nuclear magnetic resonance (NMR).

> 🔗 Photoelectricity and pair production are described in Topic 12.1.

In *X-ray medical imaging*, radiation is incident on the patient and selectively absorbed by bone and tissue. The contrast in transmitted intensity is recorded on a photographic plate or using a computer. The greater the density of the material, the more radiation is absorbed.

X-rays are produced when high-energy electrons are rapidly decelerated by a heavy metal target (typically, tungsten). Most of the kinetic energy is transferred to internal energy in the target, but a small amount is emitted in the form of X-ray photons. These photons are attenuated or scattered as they pass through the patient or other parts of the equipment.

> The **intensity** I of a monochromatic X-ray beam after attenuation is
>
> $$I = I_0 e^{-\mu x}$$
>
> where
>
> I_0 is the original intensity
>
> x is the thickness of the material
>
> μ_l is the linear absorption coefficient of the material.
>
> The unit of linear absorption coefficient is m^{-1}.
>
> The amount of attenuation is
>
> $$L_I = 10 \log_{10}\left(\frac{I_1}{I_0}\right),$$ where an initial
>
> intensity I_0 has decreased to I_1.
>
> This is similar to attenuation with optic fibres, as covered in Option C.3.

These are some of the mechanisms for attenuation and scattering.

Photoelectric effect. The photons remove inner-shell electrons from atoms. Light is emitted, as other electrons lose energy to occupy the shell. Photoelectric scattering provides contrast between tissue and bone.

Coherent scattering. This involves low-energy X-ray photons. Steps are often taken to remove these photons as they degrade image contrast.

Compton scattering. At high energies, an X-ray photon removes an outer-shell electron from an atom and a lower energy photon is emitted.

Pair production. Electron–positron pairs can be produced with very energetic photons.

Beam divergence. This causes the intensity of the beam to decrease with distance from the X-ray source.

The probability of an individual photon being absorbed or scattered is related to its chance of interacting with an atom. This is constant for a material. It is analogous to radioactive decay, and leads to a similar equation.

The linear absorption coefficient is not useful. It depends on the material state; for example, ice and steam have very different values for μ_l because their densities differ. The mass absorption coefficient μ_m depends on the absorber element and is not dependent on density.

> The equation that connects μ_l and μ_m is $\mu_m = \dfrac{\mu_l}{\rho}$, where ρ is density.
>
> The unit of **mass absorption coefficient** is $m^2 \, kg^{-1}$.
>
> **Half-thickness** is defined for a material; it is the thickness required to reduce the intensity of the X-ray beam by one-half.
>
> Again, by analogy with radioactive decay, this is given by $\mu x_{\frac{1}{2}} = \ln 2$.
>
> This leads to $x_{\frac{1}{2}} = \dfrac{\ln 2}{\mu_l} = \dfrac{\ln 2}{\rho \mu_m}$ for the two coefficients.
>
> Values of $x_{\frac{1}{2}}$ and μ_l can be obtained from graphs that show the variation of $\ln I$ with x.

Very penetrating X-rays have small absorption coefficients are *hard X-rays*, with typical wavelengths of 10 pm.

X-rays with large absorption coefficients are *soft X-rays*, with wavelengths of about 1 nm.

When the beam penetrates two or more thicknesses of different material, x_1 and x_2, with linear absorption coefficients, μ_{l1} and μ_{l2}, and when the interfaces are plane and parallel, the final intensity will be $I = I_0 e^{(\mu_{l1} x_1 + \mu_{l2} x_2)}$. There is a similar expression for mass absorption coefficients.

Many techniques are used to improve the X-ray image, whether formed on a photographic plate or by computation:

- *Filtration* by a thin metal plate to absorb low-energy photons.

- *Collimation* by lead grids above and below the patient to remove off-axis photons that blur the image.

- *Fluorescent screens* are used to improve contrast using X-ray photons that have not interacted with the photographic plate.

- Heavy elements such as *barium* are used to improve contrast in some tissues that are hard to image.

X-ray imaging is a direct and quick technique that costs far less than more sophisticated scans. However, X-rays are ionizing and represent risk to the patient and radiographer. A radiographer uses short exposure time, large distances and absorbers to minimize X-ray exposure.

Risk to the patient should be kept in perspective, though; an intercontinental flight, for example, provides a much higher dose of radiation than the average chest X-ray.

Example C.4.1

A parallel beam of X-rays is normally incident on tissue of thickness x.

The incident intensity is I_1. The intensity leaving the tissue is I_2.

The half-value thickness of the tissue is 2.5 cm.

a) Calculate the linear attenuation coefficient of the X-rays for this tissue.

b) The X-ray beam is incident on a different tissue type.

$\dfrac{I_2}{I_1}$ is smaller for the same x with the second tissue.

Compare the linear attenuation coefficient for this tissue with that in part a).

c) Explain, with reference to attenuation coefficient, why a patient drinks a liquid containing barium to help image the stomach.

Solution

a) Taking logs of the expression: $I_2 = I_1 e^{-\mu x} \Rightarrow \ln\left(\dfrac{I_2}{I_1}\right) = -\mu x$

So $\mu = \dfrac{\ln(2)}{2.5} = 0.28\,\text{cm}^{-1}$

b) The half-value thickness for the second tissue will be smaller than the first; its linear attenuation coefficient will be larger.

c) The stomach and similar tissues have large values for x and, therefore, small values for μ as the tissue does not attenuate X-rays well. Barium is a heavy metal with a high attenuation coefficient and, in liquid form, it coats the stomach wall, absorbing X-rays and showing the outline of the organ on the image.

A-scan. The equipment plots a graph of the variation with time of the reflected signal strength. The distance d of an interface from the transducer is related to time t on the scan and the speed of sound c in the tissue by $d = c \times 2t$. The factor 2 is because the wave travels to and from the reflector.

B-scan. The operator rocks the transducer from side to side to illuminate all the internal surfaces. A computer builds up an image of a slice through the patient from the resulting series of A-scans.

Ultrasound frequencies in the range 2-20 MHz can be used to image the body non-invasively.

Ultrasound is generated using the piezoelectric effect. A crystal deforms when a potential difference is applied across it. When the pd alternates at high frequency, the crystal (or ceramic material) expands and contracts at the same frequency to produce a longitudinal wave. Ultrasound obeys the normal rules for longitudinal waves and is reflected, absorbed and attenuated by matter.

To form the image (called a scan) of the internal organs of a patient, a piezoelectric transducer is placed in contact with the skin. A gel between the skin and transducer prevents significant energy loss at the air interface. A single pulse of ultrasound is transmitted by the transducer; the transmission then stops and the transducer receives pulses reflected by tissue interfaces inside the patient.

Two varieties of scan are used: the *A-scan* and the more complex *B-scan*.

Body tissues reflect and absorb ultrasounds to different extents. The *acoustic impedance* is used to compare tissues.

The **acoustic impedance** Z depends on c and the density ρ of the tissue: $Z = \rho c$.

The unit of acoustic impedance is $\text{kg}\,\text{m}^2\,\text{s}^{-1}$.

The ratio incident intensity I_0 : reflected intensity I_r is $\dfrac{I_r}{I_0} = \dfrac{(Z_2 - Z_1)^2}{(Z_2 + Z_1)^2}$

where Z_1 is the acoustic impedance of the tissue that the wave leaves and Z_2 is the acoustic impedance of the tissue that the wave enters.

The intensity reflected at an interface depends on the differences between the Z values for the two media concerned.

The operator cannot simply go to the highest frequency available because attenuation of ultrasound increases with frequency. There is a compromise between image resolution and reflected signal strength.

Other uses for ultrasound in medicine include detection of blood flow and blood speed using Doppler shift and the enhancement of blood-vessel images using microbubbles of gas.

Advantages of ultrasound techniques	Disadvantages of ultrasound techniques
• Excellent for imaging soft tissue	• Limited resolution
• Non-invasive	• Cannot transmit through bone
• Quick and inexpensive	• Lungs and digestive system cannot be imaged as the gas in them strongly reflects
• No known harmful side-effects	

Example C.4.2

Data about the velocity of sound in some materials and the density of these materials are provided.

	Sound velocity / m s^{-1}	Density / kg m^{-3}
Air	330	1.3
Gel	1400	980
Muscle	1600	1100

Demonstrate that it is necessary to use a gel between an ultrasound transmitter and the skin of a patient.

Solution

Using $Z = \rho c$, the acoustic impedances for air, gel and muscle, in kg m^{-2}s^{-1}, are

430, 1.4×10^6 and 1.8×10^6 respectively.

$\dfrac{I_r}{I_0}$ from air → muscle is $\dfrac{I_r}{I_0} = \dfrac{\left(1.4 \times 10^6 - 430\right)^2}{\left(1.4 \times 10^6 + 430\right)^2} = 0.999$

So almost all of the incident energy is reflected back to the transmitter.

$\dfrac{I_r}{I_0}$ from gel → muscle is $\dfrac{I_r}{I_0} = \dfrac{\left(1.8 \times 10^6 - 1.4 \times 10^6\right)^2}{\left(1.8 \times 10^6 + 1.4 \times 10^6\right)^2} = 0.0156$

So most of the energy is transmitted into the muscle.

Medical magnetic resonance imaging (MRI) uses nuclear magnetic resonance (NMR) to produce detailed images of parts of the body that are hydrogen (water) rich.

NMR

- Protons have charge and spin and behave as magnets.

- The spins of the protons are arranged randomly so there is no net magnetic field. Imposing a strong magnetic field aligns proton spins and individual magnetic fields line up.

- When a radio-frequency (rf) field of a particular frequency—known as the *Larmor frequency* f_L—is applied, some protons flip so their magnetic field is reversed and they enter a high-energy state. The Larmor frequency is directly proportional to the magnetic field strength (f_L/Hz $= 4.26 \times 10^7 B$, where B is in tesla).

- The protons now precess and, because their magnetic field is changing, an emf will be induced in a conductor nearby.

- When the rf field is removed, the high-energy protons revert to their low-energy state, emitting electromagnetic signals as they do so.

MRI

MRI provides positional data about origin of the NMR signals.

A magnetic gradient field is added to the original uniform field. The original field plus the gradient field are arranged so that there is a

Topic 9.4 discusses how resolution depends on wavelength.

》Assessment tip

The $\dfrac{I_r}{I_0}$ equation will be given to you in an examination as it is not provided in the data booklet.

linear variation of magnetic field strength B across the patient. The value of the Larmor frequency, therefore, also varies linearly across the patient. Specific values of f_L are associated with specific positions within the patient.

The fields are switched on to initiate proton precession and then switched off to allow the protons to relax. As they relax, the electromagnetic signal is acquired by nearby coils. The variation of the signal strength with f_L allows a computer to recover information about the number of protons emitting the signal at each point in the patient. The computer can then construct a spatial image of the proton density.

Advantages of MRI scans	Disadvantages of MRI scans
<ul><li>Images with good resolution (down to millimetres)</li><li>No exposure to radiation</li><li>No known risk from high-strength magnetic fields</li></ul>	<ul><li>Strong magnetic fields can affect heart pacemakers and metal implants can prevent a good image being produced</li><li>The rf currents can give rise to local heating in tissues</li><li>The MRI scanners are very noisy as the magnetic fields switch on and off</li><li>The space for the patient is small and can be claustrophobic</li></ul>

SAMPLE STUDENT ANSWER

a) Outline the formation of a B-scan in medical ultrasound imaging.   [3]

This answer could have achieved 2/3 marks:

> B-scan is a 2-dimensional version of A-scan. Ultrasound waves are sent to the target tissue and the reflections from the waves are used to create 1-dimensional images. When this process has repeated from different angles, a 2-D image can be formed. The ultrasound is created by vibrating crystals using AC current which is called piezoelectricity.

▲ This is an incomplete answer that correctly distinguishes between the A- and B-scans in terms of the two-dimensionality of the B variant. There is also a description of the method for obtaining the signals ('repeated from different angles').

▼ The endpoint of the process is missing. A computer is required to take the multiple reflections (a series of A-scans) and translate them into the final visual image. This use of computing is an important difference between A and B and is omitted in this answer here. There is a sentence telling the examiner how ultrasound is generated but this was not required in the question. Do not give information that is not required.

b) The attenuation values for fat and muscle at different X-ray energies are shown.   [3]

Energy of X-rays / keV	Fat attenuation coefficient / cm⁻¹	Muscle attenuation coefficient / cm⁻¹
1	2030.9767	3947.2808
5	18.4899	43.8253
10	2.3560	5.5720
20	0.4499	0.8490

A monochromatic X-ray beam of energy 20 keV and intensity I_0 penetrates 5.00 cm of fat and then 4.00 cm of muscle.

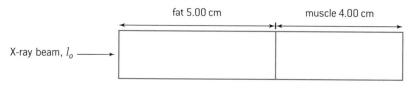

Calculate the final beam intensity that emerges from the muscle.

This answer could have achieved 2/3 marks:

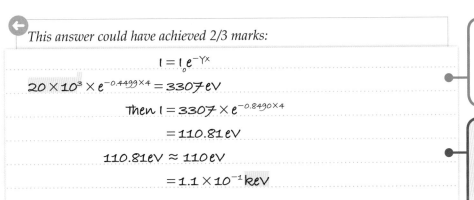

$$I = I_0 e^{-\gamma x}$$

$$20 \times 10^3 \times e^{-0.4499 \times 4} = 3307\, eV$$

Then $I = 3307 \times e^{-0.8490 \times 4}$

$$= 110.81\, eV$$

$$110.81\, eV \approx 110\, eV$$

$$= 1.1 \times 10^{-1}\, keV$$

▲ The solution is carried out in two stages and the second evaluation is correct after an error carried forward in the second line (the interim answer should be 2110).

▼ The use of 20 000 for the intensity shows a misunderstanding of what is required. The final intensity should have been quoted as a fraction of I_0. There should not have been any units quoted either.

Practice problems for Option C

Problem 1
An object is placed 12.0 cm from a diverging mirror that has a focal length of 8.0 cm.

a) Construct a scaled ray diagram for this object and mirror.

b) Estimate, using your diagram, the linear magnification of the image.

c) Comment on the advantages that a parabolic mirror has over a spherical mirror.

Problem 2
Monochromatic light from a distant point object is incident on a lens and the image is formed on the principal axis.

a) Outline the Rayleigh criterion for the resolution of two point sources by an astronomical telescope.

b) Explain why telescopes with high resolution are usually reflecting rather than refracting instruments.

Problem 3
A radio telescope has a dish diameter of 40 m. It is used to observe a wavelength of 20 cm.

Calculate the smallest distance between two point sources on the Sun that can be distinguished at this wavelength by the telescope.

Sun–Earth distance $= 1.5 \times 10^8$ km

Problem 4
The graph shows the input and output signal powers of an optical fibre.

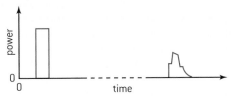

a) Explain the features of the graphs that show attenuation of the signal.

b) The width of the pulse increases with time.

   Outline reasons for this increased width.

c) Suggest, with reference to the diagram, why there is a limit of the pulse frequency that can be transmitted along a length of optical fibre.

Problem 5
a) Equal intensities of 15 keV and 30 keV X-rays are incident on a sheet of aluminium.

   15 keV X-ray half thickness $= 0.70$ mm

   30 keV X-ray half thickness $= 3.5$ mm

   Determine the ratio of the intensities of these X-ray beams after passing through an aluminium sheet of thickness 6.0 mm.

b) Explain why low-energy X-radiation is filtered out of a beam for medical use.

Problem 6
a) Draw a ray diagram to show how a converging lens is used as a magnifying glass.

b) Explain why the image cannot be formed on a screen.

c) Suggest why a magnifying glass is likely to be better in blue light than red light.

d) The converging lens is now used with an illuminated object to produce an image on a screen that is four times larger than the object. Determine the focal length of the lens.

D ASTROPHYSICS

D.1 STELLAR QUANTITIES

You must know:

- ✔ the objects in the solar system including comets, constellations, nebulae, planets and planetary systems

- ✔ definitions of a single star, binary stars, stellar clusters (open and globular)

- ✔ definition of nebulae

- ✔ definitions of galaxies, clusters of galaxies and superclusters of galaxies

- ✔ how astronomical distances are defined

- ✔ what is meant by luminosity and apparent brightness.

You should be able to:

- ✔ identify objects in the Universe

- ✔ describe the balance between gravitational force and pressure in a star

- ✔ use the astronomical unit (AU), the light year (ly) and the parsec (pc)

- ✔ describe the method of stellar parallax and comment on its limitations

- ✔ solve problems involving luminosity, apparent brightness and distance.

Example D.1.1

Outline the nature of a comet.

Solution
Comets orbit the Sun with orbital periods from years to thousands of years. Most consist of dust, rock and frozen matter. As they approach the Sun, the previously frozen material releases gases and the comet develops a gaseous tail that points away from the Sun.

The *Solar System* is a collection of objects held together by gravity. Object in the solar system include:

- the Sun, which, as a young spinning star, had a gas disc that evolved into the planets roughly 4.5 billion years ago

- the terrestrial planets Mercury, Venus, Earth and Mars

- the gas giant planets Jupiter, Saturn, Uranus and Neptune

- a Kuiper belt that consists of dwarf planets, including Pluto

- six planets that have moons

- the asteroids, which are rocky objects that orbit the Sun in a belt between Mars and Jupiter

- comets, which are irregular objects consisting of frozen materials, rock and dust; most are trapped by the Sun's gravitational field and have highly elliptical orbits with periodic times varying from years to thousands of years.

Stars form when dust and gas in a nebula is condensed by gravity. The gravitational potential energy of the material is transferred to kinetic energy and internal energy, forming a protostar. Eventually, the temperature is high enough for the nuclear fusion of hydrogen to helium to begin. Large amounts of energy are released in the form of photons.

As the energy is released, there is an outwards radiation pressure opposing the gravitational force inwards. The star is now stable. It remains stable, on the main sequence, for up to billions of years.

The processes of fusion are covered in Topic 7.2.

When the hydrogen is used up, other processes take over, and the temperature of the star (and, therefore, its colour) changes. The eventual endpoint of the star is determined by its initial mass.

Nebulae are regions of intergalactic dust and gas clouds in which stars form. Origins for nebulae are:

- gas clouds formed 380 000 years after the Big Bang when positive nuclei attracted electrons to produce hydrogen
- matter ejected from a supernova explosion.

Galaxies are collections of stars, gas and dust gravitationally bound. There are billions of stars and planets within each one. Most galaxies occur in *clusters* containing anything between dozens and thousands of galaxies. *Superclusters* of these galactic clusters account for about 90% of all galaxies and form a network of filaments and sheets. Between the network, space is apparently empty.

Parallax measurements are used to determine distances to the nearest stars. As the Earth moves across a diameter of its orbit over a six-month period, the positions of the nearest stars move relative to the 'background' of fixed distant stars. The distance across the baseline is two astronomical units (2 AU) so the parallax angle (half the six-month variation) p is related to the distance d to the star by $d = \frac{1}{p}$, where d is in parsec and p is in arc-seconds.

Stellar-parallax measurements made from the surface of the Earth allow distance estimates up to about 100 pc because turbulence in the atmosphere limits the smallest angle that can be measured. When an orbiting satellite outside the atmosphere is used, the distance measured by parallax goes up to 10 000 light years (ly).

Stars form various groupings.

Binary stars—two stars that rotate about a common centre of mass—are thought to make up about half of the stars near to us.

Stellar clusters are groups of stars held together by gravity. The number in the cluster varies from a few dozen to millions.

Open clusters are groups of a few hundred young stars with gas and dust lying between them.

Globular clusters are much older than open clusters—they were probably formed about 11 billion years ago. They have many stars and are spherically shaped as their name implies.

Constellations are groups of stars that form a pattern as seen from Earth. There is no connection between the stars, gravitational or otherwise.

The distances in astronomy are very large and involve large powers of ten. Non-SI units are frequently used to avoid this. They include the light year the astronomical unit and the parsec.

The **light year (ly)** is the distance travelled by light in one year; $1\,\text{ly} = 9.46 \times 10^{15}\,\text{m}$.

The **astronomical unit (AU)** is the average distance between the Earth and the Sun; $1\,\text{AU} = 1.50 \times 10^{11}\,\text{m}$.

The **parsec (pc)** is defined using parallax angle; a star that is 1 pc from Earth will subtend a parallax angle of 1 arc-second.

Example D.1.2

A star has a parallax angle from Earth of 0.419 arc-seconds.

a) Outline what this parallax angle means.

b) Calculate, in light years, the distance to the star.

c) State why the terrestrial parallax method can only be used for stars less than a few hundred parsecs away.

Solution

a) This is half the angle subtended by the star at the Earth over a six-month period when the Earth is at two extremes of its orbit.

b) For the parallax angle p, $d = \frac{1}{p} = \frac{1}{0.419} = 2.39\,\text{pc} \equiv 7.78\,\text{ly}$.

c) For larger distances, the parallax angle becomes small and the distortions introduced by the atmosphere produce large fractional errors in the result.

The output power of a star is known as its *luminosity L*. The star's intensity at a distance d from the star is known as its *apparent brightness b*.

» Assessment tip

The data booklet form of this equation is

$d\,\text{(parsec)} = \dfrac{1}{p\,\text{(arc-second)}}$ to

remind you of the correct units for the quantities d and p.

Luminosity L and apparent brightness b are connected by the equation $b = \dfrac{L}{4\pi d^2}$.

Knowledge of b and d allow an estimate of luminosity.

🔗 Topic 4.3 shows that the intensity I of a wave at a distance d from a source of power P is $I = \dfrac{P}{4\pi d^2}$.

🔗 Topic 8.2 shows that the power output P from a black body of temperature T is $P = e\sigma AT^4$.

Using luminosity rather than power, the $P \propto T^4$ equation gives $L = \sigma AT^4$, since $e = 1$ for a (black body) star. For a spherical star that can be treated as a black body with radius R, $L = 4\pi\sigma R^2 T^4$.

Example D.1.4

The luminosity of Antares is 98 000 times that of the Sun. Deduce $\dfrac{R}{R_\odot}$, where R is the radius of Antares and $R_\odot$ is the radius of the Sun.

Solution

Rearrange $L = 4\pi\sigma R^2 T^4$ to

give $R = \sqrt{\dfrac{L}{4\pi\sigma T^4}}$

So, $\dfrac{R}{R_\odot} = \sqrt{\dfrac{L}{L_\odot} \times \left(\dfrac{T_\odot}{T}\right)^4}$

$= \sqrt{98\,000 \times \left(\dfrac{5800}{3400}\right)^4}$

$= 910$

The radius of Antares is 900 × that of the Sun.

Example D.1.3

The apparent brightness of star X is $4.6 \times 10^{-8}\,\text{W m}^{-2}$. X has a luminosity that is 420 times that of the Sun.

Determine, in parsec, the distance of X from the Sun.

Luminosity of the Sun $= 3.8 \times 10^{26}\,\text{W}$

Solution

The luminosity of $X = (3.8 \times 10^{26} \times 420)$.

Rearranging $b = \dfrac{L}{4\pi d^2}$ gives $d = \sqrt{\dfrac{L}{4\pi b}} = \sqrt{\dfrac{3.8 \times 10^{26} \times 420}{4\pi \times 4.6 \times 10^{-8}}}$

$= 5.3 \times 10^{17}\,\text{m}$

Use $1\,\text{ly} \equiv 9.46 \times 10^{15}\,\text{m}$ to convert from m to ly: $\dfrac{5.3 \times 10^{17}}{9.46 \times 10^{15}} = 56\,\text{ly}$

Use $1\,\text{pc} \equiv 3.26\,\text{ly}$ to convert from ly to pc: $\dfrac{56}{3.26} = 17\,\text{pc}$

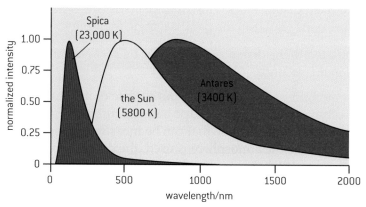

▲ **Figure D.1.1.** Normalised intensity–wavelength curves for three stars with different temperatures

Figure D.1.1 shows normalised graphs of the variation of intensity with wavelength plotted for the Sun, Spica and Antares; these graphs peak at different maximum wavelengths. This links to the Wien displacement law and gives an indication of surface temperature. The assumption that a star is a black body is reasonable, and leads to conclusions about star size and surface temperature that are explored in Option D.2.

SAMPLE STUDENT ANSWER

Alpha Centauri A and B is a binary star system in the main sequence.

	Alpha Centauri A	Alpha Centauri B
Luminosity	$1.5\,L_\odot$	$0.5\,L_\odot$
Surface temperature / K	5800	5300

a) State what is meant by a binary star system. [1]

This answer could have achieved 0/1 marks:

A binary star is a normal star which is formed from the fusion of H → He.

▼ A binary star is a system in which two stars orbit each other. The fusion process is true for all stars.

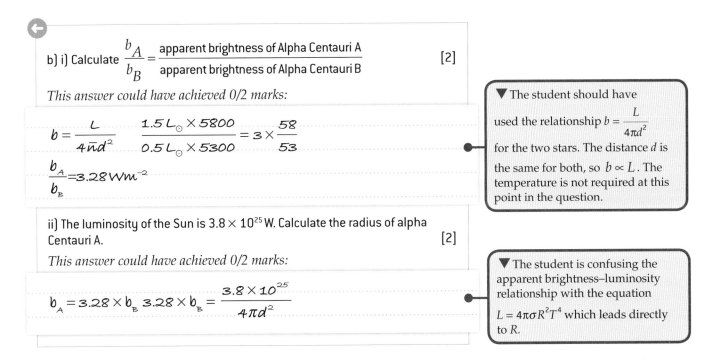

b) i) Calculate $\dfrac{b_A}{b_B} = \dfrac{\text{apparent brightness of Alpha Centauri A}}{\text{apparent brightness of Alpha Centauri B}}$ [2]

This answer could have achieved 0/2 marks:

$$b = \frac{L}{4\pi d^2} \qquad \frac{1.5\,L_\odot \times 5800}{0.5\,L_\odot \times 5300} = 3 \times \frac{58}{53}$$

$$\frac{b_A}{b_B} = 3.28\,\text{Wm}^{-2}$$

▼ The student should have used the relationship $b = \dfrac{L}{4\pi d^2}$ for the two stars. The distance d is the same for both, so $b \propto L$. The temperature is not required at this point in the question.

ii) The luminosity of the Sun is 3.8×10^{25} W. Calculate the radius of alpha Centauri A. [2]

This answer could have achieved 0/2 marks:

$$b_A = 3.28 \times b_B \quad 3.28 \times b_B = \frac{3.8 \times 10^{25}}{4\pi d^2}$$

▼ The student is confusing the apparent brightness–luminosity relationship with the equation $L = 4\pi\sigma R^2 T^4$ which leads directly to R.

D.2 STELLAR CHARACTERISTICS AND STELLAR EVOLUTION

You must know:

✔ about stellar spectra

✔ what a Hertzspring–Russell (HR) diagram shows

✔ how the HR diagram indicates stellar evolution

✔ what is meant by a Cepheid variable star

✔ the meaning of black hole, red giant, neutron star and white dwarf

✔ the mass–luminosity relationship for main sequence stars

✔ the Chandrasekhar limit for the maximum mass of a white dwarf star

✔ the Oppenheimer–Volkoff limit for the mass of a neutron star that is not to become a black hole

✔ how to describe the evolution of stars off the main sequence and the role of mass in stellar evolution.

You should be able to:

✔ explain how surface temperature of a star can be obtained from the spectrum of the star

✔ explain how stellar spectra can provide evidence for the chemical composition of stars

✔ apply the mass–luminosity relation

✔ sketch and interpret Hertzsprung–Russell (HR) diagrams

✔ identify the main regions of the HR diagram and describe the properties of stars in these regions

✔ describe the reason for the variation of Cepheid variable stars

✔ determine distance using data on Cepheid variable stars

✔ sketch and interpret evolution pathways of stars with reference to the HR diagram.

Most stars emit a continuous spectrum from their hot internal regions. This radiation passes through cooler, low-density gas in the star's outer regions which absorbs wavelengths before re-emission. Absorption lines are characteristic of the chemical elements in the cooler gas. The hottest stars often fail to show absorption lines because the hydrogen gas is completely ionized and, therefore, has no electrons to be promoted out of the ground state and absorb photons from the star's interior.

The formation of atomic emission and absorption spectra is covered in Topic 7.1.

Wien's displacement law links the surface temperature of the star and the peak wavelength:

$\lambda_{max} T = 2.9 \times 10^{-3}$, where T is the kelvin temperature of the star surface and λ_{max} is measured in metres.

Cepheid variables are **standard candles** because they allow an estimate of the distance of the variable star from Earth. The variation with pulsation period of the luminosity of Cepheid variables is known. When the pulsation period and apparent brightness b are measured, the distance to the star can be estimated.

Spectral analysis of a star is difficult because:

- there can be many elements present and the absorption lines are superimposed on each other
- Doppler broadening of the lines occurs because the atoms move
- stars often rotate; one limb of the star approaches an observer while the opposite limb moves away, which also causes Doppler shift.

Cepheid variable stars show a regular change in their emitted intensity (the light curve rises quickly to maximum luminosity and then falls slowly before the sequence repeats). These stars have moved off the main sequence into the instability strip of the HR diagram. Example D.2.1 includes an explanation of the reasons for the variation in output of the star.

Example D.2.1

Explain the periodic changes in the luminosity of a Cepheid variable.

Solution

The luminosity variation is caused by periodic expansions and contractions in the outer layers of the star. The pulsation process occurs because:

① a layer of gas in the star is pulled in by gravity

② the layer becomes compressed and more opaque because the ions are closer together

③ the temperature of the layer increases because more radiation is retained

④ the internal pressure increases and the layer is pushed outwards

⑤ as the layer expands it becomes more transparent, absorbs less radiation and cools

⑥ the layer falls inwards by gravity as the hydrostatic equilibrium between radiation pressure and gravity is disturbed.

The ①–⑥ cycle repeats.

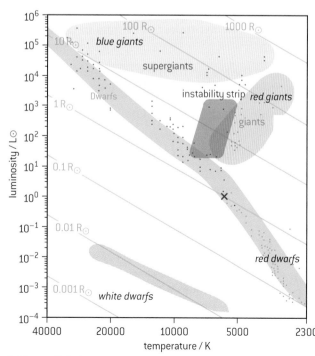

▲ Figure D.2.1. The Hertzsprung–Russell diagram

The Hertzsprung–Russell (HR) diagram represents patterns of stellar behaviour and evolution. The HR plot of *luminosity* against *temperature* (Figure D.2.1) with stars grouped according to type, shows the main features of the diagram.

The *main sequence* consists of stars producing energy by fusing hydrogen and other light nuclei. About 90% of all stars are on the main sequence. They move along it throughout their life as their luminosity and temperature change. At the bottom right of the HR diagram are small cool red stars. At the top left are large hot blue stars. The present position of the Sun is shown with an X.

Red giant stars have lower temperatures but higher luminosities than the Sun—their surface areas and diameters are much larger than the Sun.

Supergiant stars are rare, very bright and much larger than red giants. A typical supergiant emits 10^5 times the power of the Sun. About 1% of stars are red giants or supergiants.

White dwarf stars are very dense and constitute the remains of old stars. They have low luminosity and small surface area. They are cooling down and will take billions of years to do so. About 9% of all stars are white dwarfs.

The instability strip is a region where variable stars are found.

Lines of constant radius are a set of diagonal lines for constant stellar radius. The constant lines go from the upper left to the lower right of the diagram and indicate stars of the same physical size.

Stars spend different times on the main sequence. Stars with large mass and high temperature can be expected to burn their fuel more quickly than a small, cool star. This is because:

- the more massive the star, the greater the gravitational compression
- to achieve equilibrium, the radiation pressure must be greater to match this compression
- the temperature must, therefore, be greater
- this makes the rate of fusion greater as the fusion probability increases because the internal energy enables the nucleons to approach closer.

The *mass–luminosity relationship* reflects this argument.

Option D.1 outlined the processes by which fusion in a star begins. As the protostar gains mass, its temperature increases sufficiently for fusion to begin. The star joins the main sequence and remains there for as long as it has hydrogen to convert to helium.

When most of the core hydrogen is used, the star moves off the main sequence. Outward radiation pressure is no longer equal to the inward gravitational forces, and the star shrinks. Another temperature increase occurs, allowing the remaining hydrogen in the outer layers to fuse and expand so that the size of the star increases again.

The core, however, continues to shrink and heat up so that heavier elements—such as carbon and oxygen—can form by fusion. Fusion continues in the most massive stars so that they produce elements up to (the most stable) iron and nickel. From this point on, the evolution of the star depends on its mass.

> **>> Assessment tip**
>
> You need to be familiar with all the details of the HR diagram. Note, for example, the unusual axes. The temperature scale is reversed from the usual direction – it runs from high to low. Both axes are logarithmic. On Figure D.2.1, the relative intensity axis goes up in factors of 10 and the temperature axis halves every division.
>
> Examiners will expect accuracy in your sketches of HR diagrams. Make sure that the Sun's position is correct (it has a relative intensity of 1 and a temperature of about 5700 K). You do not need to draw the stars in the main sequence as a series of points; a band will do to indicate the position from about $(20\,000, 10^4)$ down to $(2500, 10^{-4})$.

> The **mass–luminosity $(M–L)$ relationship** can be written as:
>
> $$L \propto M^{3.5} \text{ or } \left(\frac{L}{L_\odot}\right) = \left(\frac{M}{M_\odot}\right)^{3.5}$$
>
> where $L_\odot$ is the luminosity of the Sun and $M_\odot$ is its mass.

Stars up to 4 solar masses	Stars greater than 4 solar masses
The core temperature is not high enough for fusion beyond carbon. As the helium becomes exhausted, the core shrinks while still radiating.	In the red-giant phase of these stars, the core is still large and at a high temperature so that nuclei fuse to create elements heavier than carbon.
Outer layers of the star are blown away as a planetary nebula.	The star ends its red-giant phase as a layered structure with elements of decreasing proton number from the centre to the outside.
Eventually, the core will have reduced to about the size of Earth and will contain carbon and oxygen ions and free electrons.	Gravitational attraction is opposed by electron degeneracy pressure, but this cannot now stabilize.
Electron degeneracy pressure prevents further shrinkage.	With a core larger than the Chandrasekhar limit, electrons and protons combine to produce neutrons and neutrinos.
	The star collapses and the neutrons rush together to approach as closely as in a nucleus.
The star is now a white dwarf with a high density ($\approx 10^9$ kg m^{-3}) and gradually cools.	The outer layers collapse inwards too, but when they meet the core they bounce outwards again forming a supernova. The effects of this are to blow the outer layers away leaving what remains of the core as a neutron star.
	Neutron degeneracy pressure opposes any gravitational collapse.
	When the mass of the neutron star is greater than the Oppenheimer–Volkoff limit, then the star will collapse gravitationally forming a black hole.

The **Chandrasekhar limit** states that the mass of a white dwarf cannot be more than 1.4 times the mass of the Sun.

The **Oppenheimer–Volkoff** limit states that there is a maximum value for the mass of a neutron star to resist gravitational collapse. The present limit is estimated to be between 1.5 and 3.0 solar masses. Neutron stars with a greater mass than the limit will form black holes.

 Black holes are discussed in more detail in Option A.5.

Electron degeneracy pressure arises because of the Pauli exclusion principle. Two electrons cannot be in identical quantum states, so the electrons provide a repulsion that counters the gravitational attraction that attempts to collapse the star.

Black holes form when large neutron stars collapse. Nothing can escape from a black hole, including photons (hence the name). Matter is attracted by, and spirals into, a black hole so that the mass of the black hole increases with time. Observations that may confirm the existence of black holes include:

- radiation emitted because, as the matter spirals it heats up, emitting X-rays

- the emission of giant jets of matter by some galaxies; it is suggested that these are caused by rotating black holes

- the modification of the trajectories of a star near a black hole by the gravitational field of the black hole.

Example D.2.2

A main sequence star X has a mass of $2.2M_\odot$. The luminosity of the Sun is $3.8 \times 10^{26}\,\text{W}$.

a) Determine the luminosity of the star.

b) i) Suggest why the time T a star spends on the main sequence is proportional to $\dfrac{M}{L}$ where M is the mass of the star.

    ii) Compare the time that X is likely to spend on the main sequence with the time that the Sun is likely to spend on the main sequence.

Solution

a) $\left(\dfrac{L}{L_\odot}\right) = \left(\dfrac{M}{M_\odot}\right)^{3.5}$ which, with a substitution, becomes

$$\left(\dfrac{L}{L_\odot}\right) = 2.2^{3.5} = 16.$$

So the star has a luminosity that is $16 \times$ that of the Sun.

b) i) The number of possible fusions depends on the initial number of hydrogen atoms and there is a fixed energy release from the fusion reaction.

    The total energy E available is proportional to the initial number of atoms and, therefore, the initial mass M.

    The luminosity L is a measure of the rate at which energy is transferred by the star.

    So the time T is given by

$$\frac{\text{total energy available for release}}{\text{rate at which energy is released}} = \frac{E}{L} \propto \frac{M}{L}$$

    ii) $\dfrac{M}{L}$ for the star $= \dfrac{2.2M_\odot}{16L_\odot} = \dfrac{M_\odot}{7.3L_\odot}$.

    The Sun is likely to spend seven times longer on the main sequence than the star.

The first graph shows the variation of apparent brightness of a Cepheid star with time.

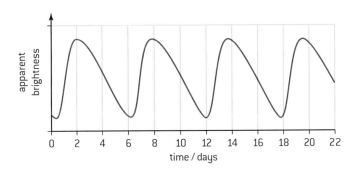

The second graph shows the average luminosity with period for Cepheid stars.

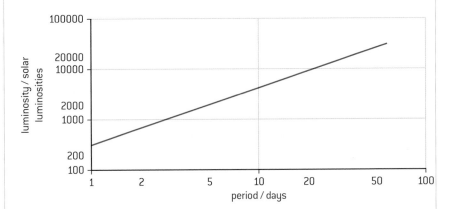

Determine the distance from Earth to the Cepheid star in parsecs.
The luminosity of the Sun is 3.8×10^{25} W.

The average apparent brightness of the Cepheid star is 1.1×10^{-9} W m^{-2}. [3]

This answer could have achieved 0/3 marks:

$$b = \frac{L}{4\pi d^2} \qquad 1.1 \times 10^{-9} = \frac{3.8 \times 10^{26}}{4\pi d^2}$$

$$d = \sqrt{\frac{3.8 \times 10^{26}}{1.38 \times 10^{-8}}} \qquad 1.1 \times 10^{-9}\ 4\pi d^2 = 3.8 \times 10^{26}$$

$$d^2 = \frac{3.8 \times 10^{26}}{1.1 \times 10^{-9} \times 4\pi}$$

$$d = 1.65 \times 10^{17}\ m = 0.07\ pc$$

▼ You often need to plan your way through a problem in Option D. Here, you need to follow the steps that professional astronomers use to determine the distance to a Cepheid variable.

Use the first graph to estimate the periodic time of the star.

Then use the second graph to find the luminosity (relative to that of the Sun) for this time period; and hence the luminosity in W.

You know now L and b so you can calculate d using $b = \dfrac{L}{4\pi d^2}$.

The answer here confuses star and Sun quantities and does not appear to use the graphs at all.

D.3 COSMOLOGY

You must know:

✔ Hubble's law

✔ how to describe the Big Bang model of the origin of the Universe

✔ what is meant by cosmic microwave background (CMB) radiation

✔ what is meant by the accelerating Universe and that redshift Z is a consequence of this

✔ the definition of cosmic scale factor R.

You should be able to:

✔ describe space and time as originating with the Big Bang

✔ describe CMB radiation and explain how it provides evidence for a Hot Big Bang model

✔ solve problems involving Hubble's law, Z and R

✔ estimate the age of the Universe assuming that the expansion rate of the Universe is constant.

Hubble's law states that the recessional speed v of a galaxy away from the Earth is directly proportional to its distance d from Earth:

$v = H_0 d$

where H_0 is the constant of proportionality, called the Hubble constant.

Galactic speeds are usually measured in $km\,s^{-1}$ with distances in Mpc; this leads to a modern value for H_0 of about $70\,km\,s^{-1}\,Mpc^{-1}$.

American astronomer Edwin Hubble compared galactic light spectra with spectra obtained in an Earth laboratory. The galactic spectra were shifted to red wavelengths as expected from the Doppler effect. However, when he used Cepheid variables to determine the galactic distances, he found that the amount of redshift depended on the distance. This led to *Hubble's law*.

Hubble showed that galaxies are moving apart. The present model of the Universe is that space and time came into existence about 13.7 billion years ago in a Hot Big Bang. At this instant, the Universe was immensely small (less than the size of an atom) and at a temperature of $10^{32}\,K$. Within one second, the Universe had cooled to $10^{10}\,K$ and was rapidly expanding. Since then, the Universe has continued to cool to its present temperature of 2.8 K.

Hubble's law allows an estimate of the age of the Universe. The steps and assumptions in estimating the Universe age are:

• Hubble's law is true for all times

• the light from the most distant galaxy has taken the age of the Universe to reach us

• the recessional speed of this galaxy from Earth is constant and (just less than) the speed of light.

The Hubble equation becomes $c = H_0(cT)$, where T is the age of the Universe and $T = \dfrac{1}{H_0}$.

With the value for H_0 of $70\,km\,s^{-1}\,Mpc^{-1}$, the estimate for the age of the Universe is about 14 billion years.

> ### » Assessment tip
>
> The speed here is a *recessional* speed, not an actual speed. Always use the term *recessional* in this context. Do not fall into the trap of imagining that the Earth is stationary. Special relativity tells us that there is no place in the Universe where absolute speed can be considered to be zero.

Example D.3.1

The variation with distance of the recessional speed of galactic clusters is plotted.

a) Identify one method for determining the distance to a galactic cluster.

b) Determine, using the graph, the age of the Universe.

Solution

a) A suitable method might be the use of a standard candle such as a Cepheid variable star. *(The use of parallax is unsuitable for distances to galaxies.)*

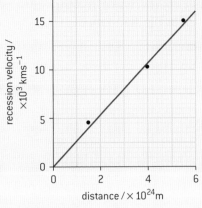
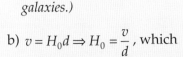

b) $v = H_0 d \Rightarrow H_0 = \dfrac{v}{d}$, which is the gradient of the graph provided.

The gradient/s^{-1} is $\dfrac{\left(16 \times 10^6 - 0\right)}{\left(6 \times 10^{24} - 0\right)} = 2.7 \times 10^{-18}$

The age of the Universe is approximately $\dfrac{1}{H_0} = 3.8 \times 10^{17}\,\text{s}$.

Further evidence for the Hot Big Bang model is the cosmic microwave background (CMB) radiation. This was predicted in the 1940s. According to the model:

- about 400 000 years after the Universe formed, the Universe temperature had fallen to 3000 K

- charged ionic matter was able to form neutral atoms with free electrons for the first time—space became transparent to electromagnetic radiation and photons could escape in all directions instead of being absorbed

- since then, the wavelength of the photons has shifted to a much longer wavelength and this black-body radiation now has an intensity peak at a wavelength of 0.07 m in the microwave region.

Satellite measurements have since confirmed that the CMB radiation exists, and that it matches the expected profile of black-body radiation. The radiation is homogeneous and almost completely isotropic (it comes from all directions and is the same from all directions). However, small variations in the CMB are now known to exist.

> **》》 Assessment tip**
>
> You should be clear about the nature of the expansion. The Universe is not expanding into anything—the fabric of space is changing. The galaxies move apart because the space between them is stretching. Electromagnetic radiation takes time to arrive from a distant galaxy and, during that time, the light wavelengths stretch so that the redshift is cosmological in origin. For this reason, take care to use the term **cosmological redshift**.

Example D.3.2

The variation of intensity with wavelength is shown for the cosmic microwave background (CMB) radiation.

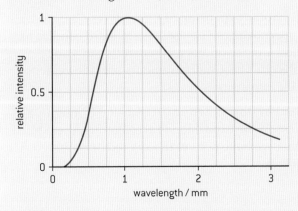

a) Estimate the temperature of the CMB radiation. State **one** assumption you make in this estimate.

b) Explain how this temperature supports the theory of the Hot Big Bang model.

Solution

a) Assume that the spectrum represents black-body radiation.

The peak wavelength is about 1.1 mm.

Using Wien's displacement law:

$$\lambda_{max} \text{ (metre)} = \frac{2.90 \times 10^{-3}}{T \text{ (kelvin)}} \quad \text{and} \quad T = \frac{2.90 \times 10^{-3}}{\lambda_{max}} = \frac{2.90 \times 10^{-3}}{1.1 \times 10^{-3}}$$

This gives a temperature of 2.6 K. As this is an estimate, 3 K will be the best way to specify the answer.

b) The model suggests that, at the time of the Big Bang, the Universe was at a very high temperature and then, with time, the Universe expanded. This meant that the wavelengths corresponding to the photon frequencies of the black-body radiation increased as the scale of the Universe increased. A long peak wavelength is equivalent to a low temperature. As the wavelengths have lengthened, the temperature has decreased to its present value which corresponds to the predictions of the model.

Redshift is a cosmological effect due to space stretching. It is not related to the relative velocities of the source and observer (unlike the standard Doppler effect). Even so, the electromagnetic Doppler equation can be used in astrophysics. The ratio of the wavelength change to the original wavelength is given the symbol z:

$$z = \frac{\Delta\lambda}{\lambda} \approx \frac{v}{c}$$

As the Universe expands, all distances are increased by the same cosmic scale factor R. In other words, light of wavelength λ_0 emitted by a galaxy will be received with a wavelength λ as the cosmic scale factor has changed from R_0 to R.

So $z = \dfrac{\Delta\lambda}{\lambda} = \dfrac{\Delta R}{R_0} = \dfrac{R}{R_0} - 1$

The existence of CMB radiation was confirmed by Penzias and Wilson when a microwave signal, picked up by their microwave antenna, proved to be CMB radiation from space. They originally thought that the signal was due to a fault in their equipment. Serendipity is sometimes very important in science.

There is now evidence that the expansion of the Universe is accelerating. Type Ia supernovae can be used as standard candles to estimate distances up to 1 Gpc; cosmological redshifts can also be determined for these objects. These measurements provide evidence for the acceleration of the expansion.

Normal mass would tend to slow down any expansion through gravitational attraction so, if the expansion is accelerating, there must be a source of energy that we have not yet discovered or observed. This has been named *dark energy*.

Example D.3.3

A galaxy has a recessional velocity relative to the Earth of 4.6×10^4 km s^{-1}.

Determine the size of the Universe relative to its present size when the light from the galaxy was emitted.

Solution

For this galaxy,

$$z \approx \frac{v}{c} = \frac{4.6 \times 10^7}{3.0 \times 10^8} = 0.153.$$

Rearranging $z = \dfrac{R}{R_0} - 1$ gives

$$R_0 = \frac{R}{(z+1)} = \frac{R}{1.153} = 0.87\,R.$$

Therefore, the size of the Universe was 87% of its present value when the galaxy emitted the light.

SAMPLE STUDENT ANSWER

Light reaching Earth from quasar 3C273 has $z = 0.16$.

a) Outline what is meant by z. [1]

This answer could have achieved 0/1 marks:

Red-shift

> ▼ z is certainly a factor that reflects the existence of cosmological redshift; however, z is not the redshift. It is the ratio of the change in wavelength due to the change in dimensions of the universe to the present value of wavelength.

b) Calculate the ratio of the size of the Universe when light was emitted by the quasar to the present size of the Universe. [1]

This answer could have achieved 1/1 marks:

$$z = \frac{R}{R_0} - 1$$

$$(0.16 + 1)(1.2 \times 10^{-15}) = R$$

$$\frac{1.3 \times 10^{-15}}{1.2 \times 10^{-15}}$$

$$R = 1.39 \times 10^{-13}$$

> ▲ The answer is correct.

> ▼ However, the solution is poorly laid out. There is no need to invent a value for R as only the ratio $\frac{R}{R_0}$ is required.

c) Calculate the distance of 3C273 from Earth using $H_0 = 68\ \mathrm{km\,s^{-1}\,Mpc^{-1}}$. [2]

This answer could have achieved 2/2 marks:

$$z = \frac{v}{c}$$

$$0.16 \times 3.00 \times 10^8 = v$$

$$v = H_0 d$$

$$\frac{0.16 \times 3.00 \times 10^8}{68000}$$

> ▲ The solution is correct and easy to follow.

d) Explain how cosmic microwave background (CMB) radiation provides support for the Hot Big Bang model. [2]

This answer could have achieved 1/2 marks::

CMB radiation provides support for the Hot Big Bang model because the microwaves are getting larger proving that the universe is expanding and the CMB are proof that at one point the universe was very very hot when it started expanding.

> ▲ The CMB radiation provides support as it is a consequence of the expansion of the Universe.

> ▼ The idea that the Universe is expanding occurs twice, but can only gain a mark once. It is easy to fall into this trap. Points in a mark scheme are awarded for separate ideas. There are also issues about what is written: the microwaves do not 'get longer'—it is their wavelength that increases. The last sentence does not give a clear view that the start of the expansion and the start of the Universe were at one and the same time.

D.4 STELLAR PROCESSES (AHL)

You must know:

✔ the Jeans criterion for star formation

✔ about the process of nucleosynthesis occurring off the main sequence

✔ the various nuclear fusion reactions that take place in stars off the main sequence including a qualitative description of the s and r processes for neutron capture

✔ about type Ia and type II supernovae and how to distinguish between them.

You should be able to:

✔ apply the Jeans criterion to the formation of a star

✔ apply the mass–luminosity equation to compare lifetimes of stars on the main sequence relative to the Sun

✔ describe the different types of nuclear fusion reactions taking place off the main sequence

✔ describe how elements heavier than iron form in stars, including details of the required temperature increases.

The **Jeans criterion** can be expressed in terms of the Jeans mass M_J below which stars cannot form, or in terms of the energies involved.

The magnitude of the gravitational potential energy of the gas cloud must be greater than the kinetic energy of the cloud for the gas cloud to collapse. In practice, a cold dense gas (small kinetic, large gravitational potential energy) will be more likely to collapse than a hot, low-density gas (high kinetic, small gravitational potential energy).

For a cold dense gas, the mass limit will be lower than for the hot diffuse cloud and the lower the mass limit, the more likely star formation will be.

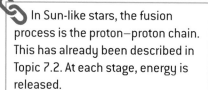

In Sun-like stars, the fusion process is the proton–proton chain. This has already been described in Topic 7.2. At each stage, energy is released.

≫ Assessment tip

These fusion processes are connected to the plot of the magnitude of binding energy per nucleon against nucleon number. Iron and nickel are at the peak of the plot and represent the most stable nucleons. Use this work in Option D as an opportunity to review and consolidate your understanding of Topic 7.2.

Stars form from dust and gas clouds that can be stable for millions of years until some event disturbs them, such as an interaction with another cloud. A large cloud then becomes unstable and collapses. As it does so, the temperature rises and, if it rises enough, nuclear fusion will begin.

For a small cloud, when collapse begins, pressure waves in the gas (travelling at the speed of sound) can cross the collapsing region quickly and restore stability. When the cloud is large, it will take too long for the waves to reach the region. Gravity will compress the gas before the sound has time to restore stability.

This is incorporated in the *Jeans criterion* for a critical mass below which a star cannot form.

In stars larger than the Sun, when the core temperature exceeds 20 MK, the CNO cycle occurs. The CNO cycle has 6 steps:

① a proton fuses with a carbon-12 nucleus to give nitrogen-13 (N-13)

② the unstable N-13 decays via positron emission into carbon-13 (C-13)

③ the C-13 fuses with a proton to give N-14

④ the N-14 fuses with a proton to give oxygen-15 (O-15)

⑤ the unstable O-15 decays via positron emission into N-15

⑥ the N-15 fuses with a proton to give C-12 emitting He-4.

The overall change is again the fusion of four protons to give He-4. The C-12 reappears at the end of the process.

The C-12 in the CNO cycle arises because the star has used up its hydrogen and its core is mostly transformed to helium. The core is shrinking as the radiation pressure falls and, therefore, it heats up. This is what causes the expansion to the red-giant phase. The increase in temperature will move the star off the main sequence.

Further nucleosynthesis is now begins.

• Two helium nuclei fuse to produce beryllium-8 (Be-8).

• Another helium fuses with the beryllium to form carbon-12.

• Another helium fuses with the carbon to form oxygen-16.

This process continues with heavier and heavier elements being produced until the most stable nuclei (iron-56 and nickel-62) are reached.

The process stops because energy must be absorbed to synthesise elements beyond nickel. The heaviest elements are created by neutron capture. Neutrons are not subject to electrostatic attraction and can approach a nucleus and be captured via the strong nuclear force. This increases the nucleon number by one but does not change the nature of the nucleus (its proton number). The new nucleus will generally emit a gamma-ray photon to de-excite. It may be unstable, allowing a neutron to decay via negative beta decay. This forms a new element.

Further neutron captures can produce increasingly heavy elements. There are two processes: the s process and the r process.

Example D.4.1

A star with a mass of the Sun moves off the main sequence.

a) Outline the nucleosynthesis processes that occur in the core of the star before and after it leaves the main sequence.

b) Outline the subsequent evolution of:

i) this star

ii) a star with a much larger mass.

Solution

a) Before leaving the main sequence, the core is fusing hydrogen to form helium. Leaving the main sequence occurs when much of the hydrogen fuel is consumed. After leaving, the core is fusing helium to form the element carbon (after a chain of fusion reactions).

b) i) The star will form a red giant. This will eventually become a planetary nebula followed by a white dwarf.

ii) A much more massive star will form a red supergiant. This will become a supernova and will then become either a neutron star or a black hole.

> **Slow neutron capture** (s process) occurs in massive stars. It leads to the production of heavy nuclides up to those of bismuth-209. Massive stars have only a small neutron flux arising from the fusion of carbon, silicon and oxygen. Unstable nuclides produced by neutron capture thus have time to decay by negative beta emission before further neutron capture occurs.
>
> **Rapid neutron capture** (r process) occurs when the new nuclei do not have time to undergo beta decay before neutron capture happens for a second time. The heavier nucleons rapidly build up, one at a time. Type II supernovae have a high neutron flux and they can produce nuclides heavier than bismuth-209 in a few minutes before beta decay has time to act. The high neutrino flux in a supernova can also lead to the creation of new elements via the weak interaction and conversion of a neutron into a proton.

The more massive the star, the shorter is its lifetime. This sounds counter-intuitive but the lifetime T of a star of mass M compared with that of the Sun (lifetime $T_\odot$ with mass $M_\odot$) is $\frac{T}{T_\odot} = \left(\frac{M}{M_\odot}\right)^{-2.5}$. A star 10 times the Sun's mass will be expected to have a lifetime of 0.3% of that of the Sun.

The reason is that the larger the star, the higher the core temperature and radiation pressure required to maintain equilibrium. The fusion rate must be higher and the hydrogen in the core is used up more quickly. The larger star spends a shorter time on the main sequence.

Supernovae are observed regularly in the night sky. They suddenly appear bright at a position where there was no brightness before. Supernovae are classified as *type I* or *type II*.

Type Ia supernovae can occur when a white dwarf in a binary star attracts mass from its companion, usually a giant star or another white dwarf. When the accreting white dwarf reaches the Chandrasekhar limit (1.4 solar masses), gravity causes it to collapse. Carbon and oxygen fuse into nickel, generating such a high radiation pressure that the star is blown apart. It reaches luminosities of around 10^{10} times that of the Sun.

Type I supernovae have no hydrogen in their line spectrum. They are the product of old stars of low mass. The type I supernovae are further classified as Ia, Ib, Ic, and so on, depending on other spectral features.

Type II supernovae have hydrogen in their line spectrum because they are young stars with large mass.

Example D.4.2

Describe how a type Ia supernova can be used to determine galactic distances.

Solution
A Type Ia supernova is a massive star in a distant galaxy that has exploded and has acquired a large luminosity that is observed on Earth. The apparent brightness of this event can be measured. The Type Ia supernovae always have the same mass when the stars explode and, therefore, all have the same luminosity. This means that they act as standard candles; the apparent brightness and this luminosity can be used to calculate the distance to the galaxy in which the supernova is located.

The reaction always occurs at this limiting mass and, therefore, we know the luminosity. A comparison with its apparent brightness provides an estimate of the distance to the galaxy that contains the supernovae.

The remnants of the supernovae expand outwards for thousands of years, eventually merging with the interstellar material to provide the material for new stars.

Type II supernovae explode through a completely different mechanism to the type 1a process.

① After all the core hydrogen has been fused, hydrogen can only fuse in a shell surrounding the (now) helium core. The core collapses to a point where the temperature will sustain helium fusion to carbon and oxygen—it takes 10^6 years to use all the helium.

② The core collapses again, allowing carbon fusion to even heavier elements with the usual temperature increase. This stage takes about 10^4 years.

③ The repetition of collapse–fusion continues with shorter time periods each time. Eventually, silicon fuses to iron-56 taking just a few days. By now there is almost no radiation pressure and gravity is the dominant force.

④ When the Chandrasekhar limit is reached, electron degeneracy pressure cannot prevent further collapse; the star implodes, producing neutrons and neutrinos leading to a neutron degeneracy pressure that resists the collapse in the form of an outwards-moving shock wave. As this wave passes through the star, taking a few hours to do so, heavy elements are formed. As the shock wave reaches the star surface, the temperature rises to $20\,000$ K and the star explodes as a supernova.

The two types of supernovae can be distinguished by observation. Type Ia have luminosities typically 10^{10} that of the Sun. Their brightness quickly reaches its maximum and then falls gradually over about six months. Type II have luminosities typically 10^9 that of the Sun. Their brightness initially peaks and then falls to a slightly lower plateau for some days, and then falls rapidly.

SAMPLE STUDENT ANSWER

a) Describe how some white dwarf stars become type Ia supernovae. [3]

This answer could have achieved 0/3 marks:

If the star has a mass of more than 3Mo, it has the chance of becoming a type Ia supernovae.

▼ The answer needs to focus on the binary nature of the star system and the way in which one of the orbiting stars accretes mass from the other until it exceeds the Chandrasekhar limit.

b) Hence, explain why a type Ia supernova is used as a standard candle. [2]

This answer could have achieved 0/2 marks:

It is used as a standard candle because of its great mass and size and due to its brightness, other stars are compared to it.

▼ There is no sense that a standard candle has a predictable luminosity and that it can be used as a standard amount of energy. When compared with the apparent brightness of an object, a distance estimation is possible.

D.5 FURTHER COSMOLOGY (AHL)

You must know:

✔ the cosmological principle and its role in models of the Universe

✔ what is meant by a rotation curve and how it relates to the mass of a galaxy

✔ what is meant by dark matter

✔ that there are fluctuations in the CMB

✔ the cosmological origin of redshift

✔ what is meant by critical density

✔ what is meant by dark energy

✔ how to sketch and interpret graphs of the variation of cosmic scale factor with time.

You should be able to:

✔ describe the cosmological principle and its role in models of the Universe

✔ describe why rotation curves are evidence for dark matter

✔ derive rotational velocity from Newtonian gravitation

✔ derive critical density from Newtonian gravitation

✔ describe and interpret the observed anisotropies in the CMB

✔ describe how the presence or otherwise of dark energy affects the value of the cosmic scale factor.

Einstein extended his general theory of relativity to cosmology by making two simplifying assumptions that have since been shown to apply on a large scale. This *cosmological principle*, together with the general theory of relativity, can be used to show that matter distorts spacetime in one of only three possible ways.

① **Positive curvature**, where travel through the Universe could lead to a return to the original position in spacetime (imagine a flat 2D spacetime world as deforming to a sphere).

② **Negative curvature**, where travel through the Universe would *never* lead to a return to the original position in spacetime (imagine a flat surface deforming to a saddle)

③ **Zero curvature** where travel through the Universe would never return (a flat 2D surface remains flat).

The appropriate model depends on the density of matter in the Universe and its value relative to a *critical density* value ρ_c that maintains zero curvature.

The theory of the Hot Big Bang model suggests that, after an initial inflationary period following the Big Bang, the expansion rate of the Universe has been decreasing. However, data from type 1a supernovae suggest that there may be an acceleration in the expansion caused by dark energy.

Therefore, it is important to know the critical density of the Universe.

Figure D.5.1 shows the variation with time of R (the cosmic scale factor from Option D.3) for these various scenarios. It includes the curve for an accelerated Universe in which the effect of the (hypothetical) dark energy exceeds that of the gravitational effects of baryonic matter and dark matter.

> The **cosmological principle** is that the Universe is:
>
> ① **homogeneous** (which means it is the same everywhere, which is true ignoring the relatively small presence of galaxies)
>
> ② **isotropic** (which means it appears the same in whichever direction we look).

> The **critical density** is given by
>
> $$\rho_c = \frac{3H_0^2}{8\pi G}$$
>
> The ratio of the actual Universe density to the critical density is given by $\Omega_0 = \frac{\rho}{\rho_c}$.
>
> • If $\Omega_0 = 1$, a flat Universe results and the Universe continues to expand to a maximum limit at a decreasing expansion rate.
>
> • If $\Omega_0 < 1$, the Universe would be open and would expand forever.
>
> • If $\Omega_0 > 1$, the Universe would be closed and expansion would eventually stop, followed by a collapse and a Big Crunch.

> A simple derivation of the critical density is given in example D.5.1. A rigorous derivation of ρ_c requires general relativity theory and is beyond the scope of the DP physics course.

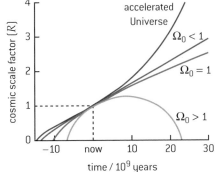

▲ Figure D.5.1. Variation of R for different density parameters

Example D.5.1

Show, by equating kinetic and gravitational potential energies, that $\rho_c = \dfrac{3H_0^2}{8\pi G}$ for the Universe.

Solution

The Universe will continue to expand providing there is enough kinetic energy to make the total energy, E_T, positive; the limiting case is $E_T = 0$.

For the Universe, $E_T = E_K + E_P$ and $E_T = \frac{1}{2}mv^2 - \dfrac{GMm}{r^2}$.

The radius of the Universe is given by $v = H_0 r$ and the mass is $\frac{4}{3}\pi r^3 \rho m$ therefore

$$\frac{1}{2}mH_0^2 r^2 = G\frac{\frac{4}{3}\pi r^3 \rho_c m}{r} \text{ and}$$

$$\rho_c = \frac{3H_0^2}{8\pi G}.$$

As space expands, the wavelength of radiation also expands. Wien's law states that $\lambda_{max}T$ is constant. There is no evidence that the shape of the black-body spectrum has changed with time so $T \propto \dfrac{1}{\lambda}$. The wavelength will scale by R together with all other dimensions and therefore $T \propto \dfrac{1}{R}$.

Evidence for the existence of dark matter comes from galactic rotation curves. Stars can lie within the bulk of the galaxy or in one of the less dense arms. Newton's law of gravitation can be used to determine the predicted rotation speed of the star for both options.

Assuming a spherical shape for the galactic hub, the star, when orbiting at distance r from the centre, will be affected by the galactic mass within its orbital radius ie $\frac{4}{3}\pi r^3 \rho$, where ρ is the galactic density.

Equating $\dfrac{GMm}{r^2}$ and $\dfrac{mv^2}{r}$ gives $v = \sqrt{\dfrac{4\pi G\rho}{3}}r$. So $v \propto r$.

For a star outside the galactic mass in a spiral arm, on the other hand,

$$v \propto \frac{1}{\sqrt{r}}.$$

These two results give a predicted graph for the variation of orbital speed with radius, as shown in Figure D.5.2.

The observed result—the rotation curve for a real galaxy—is different. Stars well outside the galaxy are moving with the same speed as those inside. A proposed explanation is that dark matter forms a shell or halo at the outer rim of the galaxy.

This material cannot be detected as it does not radiate.

Other evidence for dark matter is:

- Some galaxies in clusters orbit each other. Observations allow the masses of these galaxies to be estimated. However, the galaxies are far less bright than the mass measurements indicate.

- Radiation passing through or near massive objects is gravitationally lensed. Light from distant quasars is much more distorted than would be expected from the luminous mass contained within the galaxy. It has more mass than expected.

- Images of elliptical galaxies made in the X-ray region show halos of hot gas bound to the galaxies. The galaxies must have more mass than expected for the gas to be trapped in this way.

The ratio of visible mass to dark mass in the galaxies is thought to be $1:9$.

Candidates for dark matter, at present, are weakly interacting massive particles (WIMPs) and massive compact halo objects (MACHOs).

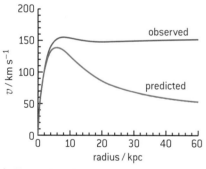

▲ Figure D.5.2. Observed versus predicted rotation curves for a galaxy

WIMPs	MACHOs
Non-baryonic subatomic particles that have different properties from ordinary matter. They must interact only weakly with normal matter and there needs to be considerable quantities of WIMP material. The theory relies on hypothetical particles not yet observed.	These include the neutron stars, black holes and small brown dwarf stars. These are high-density objects and can be detected by gravitational lensing, but they are unlikely to be present in large enough numbers to provide the amount of dark matter required throughout the Universe.

Evidence from very distant supernovae suggests that the early Universe expanded more slowly than today; this is thought to be due to dark energy. The ESA Planck mission has provided data that suggest that only 5% of the Universe is baryonic matter with 27% being dark matter and 68% of the Universe thought to be dark energy.

Some astronomers suggest that dark energy is a property of spacetime and, as space expands, so does the amount of dark energy. This could cause the expansion of the Universe to accelerate. However, this model cannot be regarded as definitive. Modification of existing theories of gravity or the development of new hypotheses may be required to incorporate these suggestions into the present Big Bang model.

Satellite images of CMB radiation over the past 20 years have revealed anisotropies—small temperature fluctuations in the background radiation. These variations appeared during cosmic inflation—a period of accelerated expansion that took place from 10^{-36} s to roughly 10^{-32} s after the Big Bang. Quantum fluctuations are thought to have occurred during this epoch and these (at that time) minute differences have become magnified into the galactic clusters observed today. The patterns seen in the satellite images are differences that were present in the radiation when the Universe became transparent at the 400 000-year mark and that have been frozen into the fabric of the Universe.

> In Option D.3 there was the suggestion that the CMB was largely isotropic. This section looks more closely at that assumption.

SAMPLE STUDENT ANSWER

The graph shows the observed orbital velocities of stars in a galaxy against their distance from the centre of the galaxy. The core of the galaxy has a radius of 4.0 kpc.

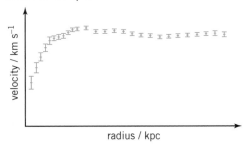

a) Calculate the rotation velocity of stars 4.0 kpc from the centre of the galaxy. The average density of the galaxy is 5.0×10^{-21} kg m^{-3}. [2]

This answer could have achieved 0/2 marks:

$$v = \sqrt{\frac{4\pi G \rho}{3}} r \qquad v = \sqrt{\frac{4\pi \times 6.67 \times 10^{-11} \times 5.0 \times 10^{-21}}{3}} \times 4$$

$$v = 3.5 \times 10^{-15} \qquad v = \sqrt{1.39 \times 10^{-30}} \times 4 \times 10^3$$

▼ The correct equation is quoted, $v = r\sqrt{\frac{4\pi G \rho}{3}}$, but there are a number of errors in its use: the 4 kpc distance is not converted to metres, there is no unit quoted for the speed, and the factor 4 kpc does not appear to have been incorporated.

b) Explain why the rotation curves are evidence for the existence of dark matter. [2]

This answer could have achieved 0/2 marks:

Rotation curves give evidence for dark matter because they show the velocities and the rotation signals emitted by the dark matter.

▼ The answer does not express the difference between predicted rotation curves (assuming no dark matter) and the observed rotation curves that can be modelled with the addition of a dark matter contribution. The fact that the predicted curves are not observed is the evidence.

D   ASTROPHYSICS

Practice problems for Option D

Problem 1

The cosmic microwave background (CMB) radiation corresponds to a temperature of 2.8 K.

a) Estimate the peak wavelength of this radiation.

b) Identify **two** other features of the CMB radiation that the Hot Big Bang model predicts.

c) The cosmic scale factor has changed by about 1000 since the emission of the CMB radiation.

Estimate, for the time when the CMB was emitted, the wavelength of a spectral line of present wavelength 21 cm.

Problem 2

The present-day value of the Hubble constant H_0 is $72 \, \text{km s}^{-1} \, \text{Mpc}^{-1}$.

a) Outline the significance of $\dfrac{1}{H_0}$.

b) A galaxy emits light of wavelength 500 nm. This light is observed on Earth to have a wavelength of 430 nm.

Deduce the distance of the galaxy from Earth.

Problem 3

a) State the Jeans criterion for the formation of a star.

b) Outline why a cold dense cloud of interstellar gas is more likely to form new stars than a hot diffuse cloud.

Problem 4

Iron is one of the most stable elements in terms of its nuclear instability.

Explain how neutron capture can lead to the production of elements with proton numbers greater than that of iron.

Problem 5

a) Describe two characteristics of a red supergiant star.

b) Explain what is meant by a constellation.

Problem 6

The luminosity of star X is 98 000 times that of the Sun and it has an apparent brightness of $1.1 \times 10^{-7} \, \text{W m}^{-2}$.

Luminosity of Sun = $3.9 \times 10^{26} \, \text{W}$

a) Calculate, in pc, the distance of X from Earth.

b) State an appropriate method for measuring the distance of X from Earth.

c) Another star Y has a luminosity that is the same as X and is on the main sequence.

Deduce the mass of star Y relative to the mass of the Sun $M_\odot$.

d) Suggest the subsequent evolution of star Y.

220

INTERNAL ASSESSMENT

During your course you will do an internal assessment project (IA). This project takes around 10 hours and is an integral part of the assessment. It accounts for 20% of your total mark in the DP physics course.

The type of project is flexible. It might be:

- experimental data collection from laboratory work
- database investigation and analysis
- spreadsheet processing with data you have collected
- the investigation of an experimental simulation.

Collaboration between students is allowed in the early stages of the IA when data are being collected. However, you must carry out the data analysis independently.

To ensure that you earn as many of the available marks as possible, you should pay particular attention to the assessment criteria, given in the following table.

Criterion	Weighting
Personal engagement	8%
Exploration	25%
Analysis	25%
Evaluation	25%
Communication	17%

You may be allowed to see examples of IA work produced in your school in previous years. If so, look carefully at the quality of work that gained high marks and, in contrast, identify why students who gained low marks scored poorly. Remember that you are seeing the final report after two weeks or so of work and some of the developmental thinking may be missing. However, a good report should reflect the refinements in both thinking and methodology that occurred as the project progressed.

Choosing a topic and personal engagement

You need to show that you have a high level of engagement with your IA. You may show creativity in your thinking, or express scientific concepts in a personal and engaged way.

If you choose a project that is based on a personal interest, then you will be more likely to succeed. For example, if you are a woodwind player, you might investigate some aspect of the physics of this instrument. The key is to think creatively. Sometimes, the aspects can be developed in laboratory work; alternatively, there may be suitable data on the Internet or published in books. Look at what others have done, but do not use anyone else's work (words or data) without giving full credit to the author.

Where possible, try to carry out preliminary experiments before the IA time begins. This preliminary work should appear with evaluative comments in your final report.

> ## >> Assessment tip
>
> When writing your introduction, ensure that
>
> ✔ your personal engagement is clear
> ✔ a research question is given that is meaningful, relevant and achievable.
> ✔ you use technical language
> ✔ you include any useful diagrams
> ✔ you do not assume any specialist prior knowledge and any symbols given are defined.

Try to make sure that your personal engagement pervades all your work and that this engagement is clear throughout the report.

Exploration

You need to decide on a **research question**. It must be focused and identify all the issues linked to it. The physics concepts and techniques should arise naturally from the topic and not be forced into it.

When you have chosen your topic, make sure your question is well defined.

▼ **Bad question**

What happens when spheres drop through water?

▲ **Good question**

How does the terminal speed of a sphere vary with temperature?

Your question should clearly identify what you will be investigating, with the independent and dependent variable expressed.

Ask yourself the following questions about what you propose to do.

- Is the physics of an appropriate standard (in other words, is it too easy or too hard)?

- Can you achieve your plans in the time available and still leave time for a thorough analysis?

- Is there scope for refinement of the IA as it progresses?

- Are your proposed experiments likely to lead to firm conclusions?

- Will you be as engaged with the project and its outcomes on the last day as on the first?

If you can answer 'yes' to all these, then your research task is suitable for you.

Planning an IA involves much more than just choosing the apparatus and the experimental methodology. You should make predictions from the outset and modify these in the light of your work.

If you choose an experimental topic, ensure that your research question will allow the collection of appropriate data. Will it allow you to access a continuously variable dependent variable with a sensible independent variable? Can your variables be made to vary over a wide range so that there is a significant difference between the lowest and highest values?

>> **Assessment tip**

An **independent variable** is one that is changed and manipulated by you.

A **dependent variable** is the one that you will measure or observe.

Control variables are the ones that are held constant during the measurement.

Continuous variables are ones that can take an infinite number of values (such as an ammeter reading); **discrete variables** can only take a fixed number (such as surfaces on which a block slides: glass, sandpaper, concrete, and so on).

Here are some things you might consider.

- What are your independent variable(s) and your dependent variable(s)? What will you keep constant and how will you achieve this?

- Are the variables continuous or discrete?

- How will you vary the independent variable and measure it? How will you measure the dependent variable?

- Is the experimental method safe in all respects? Can safety be improved?

- What do you predict as the outcome? Is this informed by a sensible theory?

- What graphs will help you to know that the prediction is correct/incorrect? Notice that a negative result is just as significant as one that confirms your prediction. An IA does not "fail" because the research question has not been confirmed.

>> **Assessment tip**

These can be summarised as follows.

✔ Make a prediction and know what graphs/display techniques will demonstrate it
✔ Make measurements safely and display them clearly and effectively
✔ Manipulate data accurately and use them to verify/deny your prediction, leading to a new refined prediction.

This cycle is sometimes known as the scientific method:

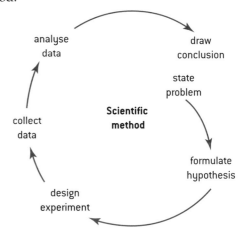

Here are some guidelines that you should consider as you work through your project.

- Keep accurate *daily* notes of your work including all references as you find them. Many working scientists use hard-backed notebooks but if you go for an electronic record on a computer or tablet, back up your work at the end of every session.

- Spend some time during every work session reviewing the work you have done. Carry out the analysis of the day's data, plot any graphs from these data, and plan out the objectives and targets for the following session. It may be appropriate to do these analyses at home as laboratory time during a project is precious. Remember: IB learners are reflective.

>> **Assessment tip**

When analysing your data:

✔ be careful to compare the outcome with the original prediction

✔ if appropriate, make use of a spreadsheet program for analysing the data and offering a possible equation to model them.

Analysis

Data from any experiment or investigation need to be processed, analysed and interpreted. The analysis criterion focuses on this aspect. The research question must lead to a detailed and valid answer. There must be enough data to justify your conclusion and for you to form a view of the experimental uncertainties in your data. In essence, the analysis must be directed towards answering the research question.

Graphs with a curved trend show the basics of how one variable varies with the change in another.

Straight-line graphs give more information than this, including the exact nature of the relationship. Manipulate your data, if at all possible, to produce a straight trend.

Look beyond the obvious and see if there are any hidden trends.

It is important to undertake a full analysis of the errors and uncertainties in your data. Remember that errors can be random or systematic (or both in the same measurement). Look carefully at the apparatus and the data to decide on the magnitude of these errors. Make realistic estimates of them where possible.

Evaluation

It is important to evaluate an experiment, whether you are a student or a professional scientist. Consider the quality of the work (experimental or analytical) and any issues that arose as you completed it. If possible, compare your results with accepted values or published work of other scientists. If you recognise problems in your work, discuss these and suggest how they could be overcome (if you were able to repeat your work). Scientists should seek to identify any shortcomings in their work; the ability to do this is a strength not a weakness.

The internal assessment in Group 4 is about your ability to carry a short project through from beginning to end rather than your ability to invent new science; that is unlikely to happen. You should provide your teacher with evidence of your progress in the project.

>> **Assessment tip**

Here are some possible transformations to allow you to obtain a straight trend.

Predicted relationship	Plot as		Equation	Gradient	Intercept on y-axis	Notes
	y-axis	x-axis				
$y = kx + c$	y	x	$y = mx + c$	k	c	
$y = kx^2$	$\sqrt{y}$	x	$\sqrt{y} = x\sqrt{k}$	$\sqrt{k}$	0	This reduces the impact of errors in x
$y = kx^2 + c$	y	x^2		k	c	
$xy = c$	y	$\dfrac{1}{x}$	$y = \dfrac{c}{x}$	c	0	
$y = x^n$	$\ln y$	$\ln x$	$\ln y = n\ln x$	n	n/a	HL only
$y = Ae^{kx}$	$\ln y$	x	$\ln y = \ln A + kx$	k	$\ln A$	HL only

- Write down everything *day-by-day* (you can edit it later).

- Consider producing a rough project draft at about three-quarters of the way through the work. This will allow you to identify simple improvements when you still have time to implement them.

- Give details of preliminary work and show how your ideas developed.

- Make sure that your comments are aligned with your data and graph plots.

- Consider safety issues and whether they were adequate.

- Draw attention to anomalies in the results and, if possible, the reasons for them.

- Discuss the extent to which your results support your conclusion, which should be strongly linked to your research question.

Finally, be critical of your own work. There is no such thing as perfect science. A hallmark of a good scientist is the ability to be self-critical.

▼ **Bad evaluation**

Simple statements that just repeat the trend or state that it is hard to draw a conclusion.

Description of practical issues that had to be overcome, rather than consideration of fundamental issues of the experimental method.

▲ **Good evaluation**

Focuses on the extent to which the research question can be answered.

Discussion of the limitations of the experimental method, and realistic suggestions for improving the method.

Identifies any preliminary work that could have been carried out.

>> **Assessment tip**

Your evaluation must focus strongly on the research question and the extent to which you feel you have been able to answer it.

Communication

Scientific writing should be concise and effective. Aim for the highest standard of presentation that you can achieve. The main part of the text should be about 6–12 pages. Large quantities of data can be placed in an appendix—do not allow them to intrude into the main text as they will disrupt the flow that you want to achieve.

A report should contain:

- a statement to put the work in context

- the research question

- an outline of the physics underlying the IA

- details of any preliminary work

- a complete account of the work you did; use an appropriate format for this work, which may depend on the nature of your data collection

- full results (these may need to be in an appendix) and a complete analysis of the results that is linked to the physics quoted earlier

- a conclusion that presents the findings clearly and unambiguously together with your evaluation and reflections on the whole investigation.

>> **Assessment tip**

Diagrams can and should be used where appropriate. Any diagrams should be fully annotated.

Referencing

You are allowed and encouraged to use secondary sources in your report, but you must make sure you correctly reference any work that is not your own. It is academic theft to take someone else's work and pass it off as your own.

Although the IB does not recommend a particular reference style, a common method for citing scientific material is the Harvard style.

In the main text you add a citation using the format: (Homer, 2019, p. 22). This reference cites this particular page in this book.

At the end of your report you must give all the references together.

Web references are similar; give the name of the author and the date if possible, and also the date on which you accessed the reference, for example, (Garza, 2014).

For scientific papers and web pages the formats in the *References* are as follows:

REFERENCES

Homer, D R 2019, *IB Prepared Physics*, Oxford University Press, Oxford.

Garza, Celina 2014, *Academic honesty – principles to practice*, International Baccalaureate Organisation, viewed 15 October 2018

<https://www.ibo.org/contentassets/71f2f66b529f48a8a61223070887373a/academic-honesty.-principles-into-practice---celina-garza.pdf>

Checklist

Here is a checklist that you may wish to use to check that you are addressing all of the marking criteria.

Personal engagement	
Is there evidence that I have linked the IA to my personal engagement?	Is this evidence **clear** and does it show significant **thought, initiative** and **creativity?**
Have I justified my research question and the topic in general?	Has my work shown **personal significance** and **interest**?
Have I shown a personal input and initiative?	Is this evident in the **design** phase, the **implementation** and the **presentation** of my work?

Exploration	
Have I identified my topic and research question?	Is my description **relevant** and fully **focused**?
Have I provided background information?	Is this entirely **appropriate**, **concise** and **relevant** and does it enhance a reader's understanding of the context
Have I made my experimental method clear?	Is the method **highly appropriate** to the investigation?
Have I reported the factors that influenced my data collection?	Have I discussed the **relevance** and **reliability** of my data and whether there are **enough data**?
Have I commented on other relevant issues that have an impact on the investigation?	Have I discussed the **safety** of my work for myself and others? If necessary, have I considered the **ethical** and **environmental** impacts of my work?

Analysis	
Have I included my raw data either in the main text or in an appendix?	Are my data sufficient to reach a **detailed** and **valid** conclusion?
Is my data processing reported appropriately?	Have I **checked my data**? Are they **appropriate** and sufficiently **accurate** for me to address my research question in a consistent way?
Have I considered uncertainties?	Is my analysis of error **consistent**, **full** and **appropriate**?
Have I provided an interpretation of my processed data?	Are my conclusions **correct**, **valid** and in enough **detail** for the reader?

Evaluation	
Does my conclusion relate to my data?	Have I **described** and **justified** the conclusion in enough detail? Do my data **support** my research question?
Does my conclusion relate to my scientific context as summarised earlier in the report?	Have I **justified** my investigation in the context of the accepted science using a relevant approach?
Have I considered the strengths and weaknesses of my investigation?	Have I discussed issues such as **data limitations** and **error sources** and shown that I understand their relevance to my conclusion
Have I suggested further work arising directly from my IA?	Have I said how I would **extend** the work is I had more time? What **realistic** and **relevant improvements** would I make if I were doing the IA again?

Communication	
Have I given thought in my report to:	
• presentation	Does my presentation help to make my work clear to a reader? Have I eliminated as many errors as possible?
• structure	Is my report **clear** and **well structured**? Is the information on my focus, what I did, and how I analysed my data coherent and well linked? Does my personal engagement pervade the report and is it a genuine engagement?
• relevance	Is my report **concise** and **relevant** so that a reader can easily understand my focus, my analysis and my conclusions?
• technical language?	Have I used all my **physics terminology** correctly and appropriately?

PRACTICE EXAM PAPERS

At this point, you will have re-familiarized yourself with the content from the topics and options of the IB Physics syllabus. Additionally, you will have picked up some key techniques and skills to refine your exam approach. It is now time to put these skills to the test; in this section you will find practice examination papers, 1, 2 and 3, with the same structure as the external assessment you will complete at the end of the DP course. Answers to these papers are available at **www.oxfordsecondary.com/ib-prepared-support**.

Paper 1

SL: 45 minutes

HL: 1 hour

Instructions to candidates

- Answer all the questions.

- For each question, choose the answer you consider to be the best and indicate your choice on the answer sheet (provided at **www.oxfordsecondary.com/ib-prepared-support**).

- A clean copy of the **physics data booklet** is required for this paper.

- The maximum mark for the SL examination paper is **[30 marks]**.

- The maximum mark for the HL examination paper is **[40 marks]**.

SL candidates: answer questions 1–30 **only**.

HL candidates: answer **all** *questions.*

1. The length of a rectangle is 2.34 cm and the width is 5.6 cm.

   What is the area of the rectangle to an appropriate number of significant figures?

   A. 1×10^1 cm^2
   B. 13 cm^2
   C. 13.1 cm^2
   D. 13.10 cm^2

2. Which list contains **only** fundamental (base) units?

   A. kilogram, mole, kelvin
   B. kilogram, coulomb, ampere
   C. ampere, mole, volt
   D. coulomb, mole, Celsius

3. The acceleration of an object varies linearly from zero to 30 m s^{-2} in 20 s.

   What is the change in speed of the object after 20 s?

   A. 1.5 m s^{-1} C. 150 m s^{-1}
   B. 3.0 m s^{-1} D. 300 m s^{-1}

4. Mass P slides along a horizontal surface and collides with an identical mass Q. The speed of P immediately before the collision is $+u$.

   What are the velocities of P and Q immediately after the collision?

	Velocity of P	Velocity of Q
A.	$-\dfrac{u}{2}$	$+\dfrac{u}{2}$
B.	$-u$	$+u$
C.	0	$+u$
D.	$-u$	0

5. A boy fires a pellet of mass m a vertical distance d into the air using a rubber band. The time from the instant the boy releases the pellet until it leaves the rubber band is Δt.

   What is the power developed by the rubber band as it fires the pellet?

   A. $mgd\Delta t$ C. $\dfrac{mg\Delta t}{d}$

   B. $\dfrac{mgd}{\Delta t}$ D. $\dfrac{mg}{d\Delta t}$

6. A constant force acts on a mass. What describes the variation of kinetic energy of the mass with the work W done on the mass?

   A. $E_{ke} = $ constant
   B. $E_{ke} \propto W$
   C. $E_{ke} \propto W^2$
   D. $E_{ke} \propto W^{-2}$

7. A boy throws a pebble horizontally with speed v across horizontal ground. The pebble strikes the ground a horizontal distance d from the point of projection a time t after it was released.

   He now throws an identical pebble from the same starting height with an initial speed $2v$.

   What are the time after release and the horizontal distance travelled when the second pebble hits the ground?

	Time after release	Distance travelled
A.	t	d
B.	$2t$	$2d$
C.	t	$2d$
D.	$2t$	d

8. For a body, Newton's second law states that:
   A. change of momentum $\propto$ external force acting
   B. force acting = acceleration
   C. change of momentum per unit time = external force acting
   D. force acting $\propto$ mass

9. An object on a planet drops from rest through a distance of 4.0 m in a time of 1.0 s. Air resistance is negligible.

   What is the acceleration of free fall?

   A. $2.0\,\mathrm{m\,s^{-2}}$
   B. $4.0\,\mathrm{m\,s^{-2}}$
   C. $8.0\,\mathrm{m\,s^{-2}}$
   D. $12.0\,\mathrm{m\,s^{-2}}$

10. An electric heater transfers a power P to a liquid of mass m. In a time t the temperature of the liquid changes by ΔT.

    What is the specific heat capacity of the liquid?

    A. $\dfrac{Pt}{m\Delta T}$
    B. $\dfrac{Pm}{t\Delta T}$
    C. $\dfrac{m\Delta T}{Pt}$
    D. $\dfrac{P}{m\Delta T}$

11. A fixed mass of an ideal gas has pressure p, density ρ, temperature T and volume V.

    What is always proportional to p?

    A. V
    B. V, ρ
    C. V, T
    D. ρ, T

12. What is proportional to the temperature of an ideal gas?

    A. frequency of collision with other molecules
    B. mean kinetic energy of the molecules
    C. total momentum of the molecules
    D. mean velocity of the molecules

13. A system executes simple harmonic motion.

    What is proportional to the restoring force that acts on the system?

    A. displacement from equilibrium
    B. amplitude of the oscillation
    C. kinetic energy of the system
    D. time period of the oscillation

14. In a standing wave:

    A. the kinetic energy of the wave is proportional to the amplitude of oscillation of the medium
    B. each part of the wave oscillates with a different frequency and phase from every other part
    C. the amplitude along the wave varies with time
    D. no energy is transferred along the wave

15. A transverse wave travels along a string. Two points on a string are separated by half a wavelength.

    Which statement about the displacements of these points is **true**?

    A. they are constant
    B. they are the same as each other
    C. they are $\dfrac{\pi}{2}$ rad out of phase with each other
    D. they are π out of phase with each other

16. A pipe is open at one end and closed at the other. The length of the pipe is D.

    What is the wavelength of the third harmonic standing wave for this pipe?

    A. $\dfrac{3D}{2}$
    B. $\dfrac{2D}{3}$
    C. $3D$
    D. $\dfrac{D}{3}$

17. A beam of monochromatic light is incident on a single slit of width b. A pattern forms on a screen a distance D from the slit.

What changes in b and D, carried out separately, lead to an increase in the width of the central maximum of the pattern?

	Change in b	Change in D
A.	increase	increase
B.	increase	decrease
C.	decrease	increase
D.	decrease	decrease

18. What is the definition of *electromotive force* of a device?

A. power supplied by the device per unit current

B. force that the device provides to drive charge carriers in the circuit

C. energy supplied by the device per unit current

D. electric field acting on a charge moving in the device

19. Three identical lamps X, Y and Z are connected to a cell of emf 6.0 V and negligible resistance so that all components are in parallel. The lamps are labelled '6.0 V, 1.2 W'.

Lamp Y breaks and does not conduct current.

What happens to lamps X and Z?

A. the current in them increases

B. the current in them stays the same

C. the current in them decreases

D. the lamps immediately burn out

20. Electrons in a horizontal beam are moving due north. What is the direction of the magnetic field due to the electron beam vertically above it?

A. to the east

B. to the west

C. upwards

D. downwards

21. Two wires, A and B, have a circular cross-section and identical lengths and resistances. Wire A has twice the diameter of wire B.

What is $\dfrac{\text{resistivity of wire A}}{\text{resistivity of wire B}}$?

  A. 0.25   B. 0.5   C. 2   D. 4

The following information is needed for questions 22 and 23. An object is suspended by a string from a fixed point on a ceiling. The object is made to move at a constant speed on a circular horizontal path. The object has a mass M with a path radius of R and makes one complete revolution in time T.

22. What is the direction of the resultant force that acts on the object?

A. outwards from the centre of the circular path

B. upwards along the string

C. towards the centre of the circular path

D. along the velocity vector of the mass

23. What is the net force acting on the mass?

A. $4\pi^2 Tmr$

B. $\dfrac{4\pi^2 mr}{T}$

C. $\dfrac{4\pi^2 mr}{T^2}$

D. $4\pi^2 T^2 mr$

24. Which pair are both classes of hadrons?

A. leptons, bosons

B. mesons, baryons

C. mesons, leptons

D. bosons, mesons

25. What is true for the proton (Z), neutron (N) and nucleon (A) numbers?

A. $Z = A - N$

B. $A = Z - N$

C. $A = N - Z$

D. $Z = A + N$

26. What evidence is provided by the Rutherford–Geiger–Marsden experiment?

A. alpha particles have discrete amounts of energy

B. the positive charge of an atom is concentrated in a small volume

C. nuclei contain protons and neutrons

D. gold atoms have large magnitudes for the binding energy per nucleon

27. A radioacitve nuclide with a half-life of 1 hour has an initial activity A_0 at time $t = 0$.

When:

- $t = 1$ hour, the activity is A_1
- $t = 2$ hours, the activity is A_2
- $t = 3$ hours the activity is A_3.

What is $\dfrac{A_0}{A_2}$?

A. $\dfrac{A_1}{A_3}$ B. $\dfrac{A_0}{A_3}$ C. $\dfrac{A_0}{2}$ D. $\dfrac{A_0}{3}$

28. What are the overall energy transformations in solar heating panels and photovoltaic cells?

	Solar heating panels	Photovoltaic cells
A.	solar to thermal	solar to thermal
B.	solar to electrical	solar to thermal
C.	solar to thermal	solar to electrical
D.	solar to electrical	solar to electrical

29. A black body emits radiation that has a peak wavelength at λ_p at a maximum intensity I_p. The temperature of the black body is then increased.

What are the changes to the peak wavelength and the maximum intensity?

	Change to peak wavelength	Change to intensity
A.	greater than λ_p	greater than I_p
B.	less than λ_p	greater than I_p
C.	greater than λ_p	less than I_p
D.	less than λ_p	less than I_p

30. A power station burns fuel with an overall efficiency of E. The energy density of the fuel is D and its specific energy is S; the rate at which the mass of fuel is consumed is M.

What is the power output of the power station?

A. $M \times S \times E$

B. $M \times D \times E$

C. $\dfrac{M \times S}{E}$

D. $\dfrac{M \times D}{E}$

*The following questions are for HL candidates **only**.*

31. Two astronomical objects have an angular separation at the Earth of 0.50 mrad. Their images are just resolved in the 2.5 cm wavelength with a radio telescope.

What is the diameter of the circular telescope aperture?

A. 6 km   **B.** 600 m   **C.** 60 m   **D.** 6 m

32. The wavelength of a spectral line measured on Earth is 500 nm. The same line is observed from an astronomical source moving away from Earth at $\dfrac{c}{10}$.

What is the wavelength from the source when measured on Earth?

A. 50 nm

B. 450 nm

C. 550 nm

D. 5000 nm

33. An object performs simple harmonic motion with time period T and amplitude x_0. The displacement of the object is at a maximum at time $t = 0$.

What is the displacement when $t = \dfrac{T}{4}$?

A. $-x_0$

B. 0

C. $\dfrac{x_0}{2}$

D. x_0

34. Two isolated spheres have the same gravitational potential at their surfaces.

What is identical for the two spheres?

A. $\dfrac{\text{radius}^3}{\text{mass}}$

B. $\dfrac{\text{radius}^2}{\text{mass}}$

C. $\dfrac{\text{radius}}{\text{mass}}$

D. radius

35. Two parallel metal plates X and Y are separated by a distance s in a vacuum. X is positively charged and Y is negatively charged; both plates have the same magnitude of charge. A particle with charge q is accelerated from rest at X to Y gaining kinetic energy E_{ke} as it reaches Y.

What is the magnitude of the electric field strength between X and Y?

A. $\dfrac{E_{ke}}{qs}$

B. $\dfrac{sE_{ke}}{q}$

C. $\dfrac{qs}{E_{ke}}$

D. $\dfrac{q}{sE_{ke}}$

36. A transformer has an efficiency of 100%. The primary coil has N_p turns and a power P is input to it. The number of turns on the secondary coil is N_s.

What is the power output by the transformer?

A. $\dfrac{N_s}{N_p}P$

B. $\dfrac{N_p}{N_s}P$

C. P

D. $\dfrac{1}{P}$

37. A 0.10 mF capacitor is initially uncharged. Charge is made to flow onto the capacitor at a constant rate of $10\,\mu C\,s^{-1}$. The dielectric in the capacitor breaks down when the potential difference across the capacitor plates is 10 kV.

What is the charging time before the capacitor breaks down?

A. 10^{-8} s   **B.** 10^{-5} s   **C.** 10^2 s   **D.** 10^5 s

38. A coil is connected to an ammeter. A bar magnet is placed along the axis of the coil.

The magnet and coil are moved as follows.

① The magnet and coil are moved with the same velocity.

② The magnet is moved towards the stationary coil.

③ The coil is moved towards the stationary magnet.

In which situations will a current be indicated on the ammeter?

A. ① and ②
B. ① and ③
C. ② and ③
D. ①, ② and ③

39. An electron is accelerated from rest through a potential difference V. The de Broglie wavelength of this electron is λ.

What is the de Broglie wavelength when the electron is accelerated through 6 V?

A. $\dfrac{\lambda}{6}$ **B.** $\dfrac{\lambda}{3}$ **C.** 3λ **D.** 6λ

40. An electron neutrino is emitted during an interaction.

What is the charge on an electron neutrino and what is the nature of the interaction?

	Charge on electron neutrino	Interaction
A.	0	During β^+ emission
B.	0	During β^- emission
C.	$+e$	During β^+ emission
D.	$+e$	During β^- emission

Paper 2

Instructions to candidates

- Answers must be written within the answer boxes (answer sheets are provided at **www.oxfordsecondary.com/ib-prepared-support**).
- A calculator is required for this paper.
- A clean copy of the **physics data booklet** is required for this paper.
- The maximum mark for the SL examination paper is **[50 marks]**.
- The maximum mark for the HL examination paper is **[90 marks]**.

*SL candidates: answer questions 1–5 **only**.*

*HL candidates: answer **all** questions.*

1. An object is released, from rest, in air above the Earth's surface.

   The graph shows the variation of the speed v of the object with time t. Air resistance is not negligible.

   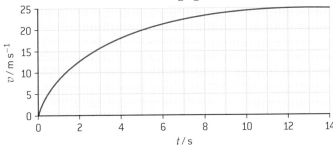

   a) Explain how the graph indicates that the air resistance acting on the object increases with time. [3]

   b) Outline, with reference to the forces acting, why the object reaches a terminal speed. [3]

   c) Estimate the distance from its starting point at which the object reaches its terminal speed. [3]

   d) The object, of weight 15 N, falls through 350 m. It has a specific heat capacity of $330\,\mathrm{J\,kg^{-1}\,K^{-1}}$.

      i) Show that the total energy of the object has decreased by about 4.8 kJ. [2]

      ii) Estimate the increase in temperature of the object. [2]

      iii) Explain one assumption that is made in part d) ii). [1]

*Question 1(e) is for HL candidates **only**.*

   e) The object falls to the Earth's surface and, during the impact, forms a crater 12 cm deep.
      Calculate the average force that acts on the object as it is brought to rest. [3]

2. a) Outline **two** ways in which a standing wave differs from a travelling wave. [2]

   b) A string of length 1.8 m is fixed at both ends. The string is made to vibrate in its third-harmonic mode.

      i) Draw the standing wave produced. Label all the nodes present. [2]

      ii) Explain how a standing wave forms on a string fixed at both ends. [3]

iii) When the frequency is increased by 23 Hz, the fourth-harmonic standing wave forms on the string.

Determine the speed of the wave on the string. [3]

3. Two conductors, A and B, have potential difference–current (V–I) characteristics as shown.

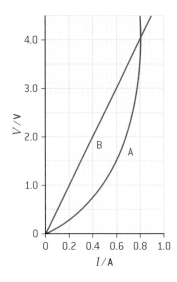

a) Sketch a graph to show the variation of the resistance of A with potential difference across it. No numbers are required on the axes of your graph. [2]

b) A and B are connected in a series circuit with a battery of emf ε and negligible resistance. The energy transferred each second in A and B is the same.

i) Determine ε. [4]

ii) Calculate the total power transferred in the circuit. [2]

*Question 3(c) is for HL candidates **only**.*

c) Outline whether A or B is an ohmic conductor. [2]

4. a) Outline what is meant by the *binding energy* of a nucleus. [2]

b) One possible reaction when plutonium $\left(^{239}_{94}\text{Pu}\right)$ undergoes nuclear fission is

$$^{239}_{94}\text{Pu} + ^{1}_{0}\text{n} \rightarrow ^{91}_{38}\text{Sr} + ^{146}_{56}\text{Ba} + x^{1}_{0}\text{n}$$

i) Calculate x. [1]

ii) Determine the binding energy per nucleon for plutonium.

Mass of plutonium nucleus $= 239.052157\,\text{u}$ [3]

c) Explain, with reference to nuclear forces, why plutonium has more neutrons than protons in the nucleus. [3]

5. A fossil-fuel power station burns coal. Its energy is transferred to consumers.

a) Outline what is meant by the *specific energy* of the fossil fuel. [1]

b) Describe the energy transfers that take place in the turbine and dynamo of the power station. [3]

c) A fossil-fuel power station has a maximum power output of 3.5 GW.
A nuclear power station is to provide the same maximum power output throughout the year as that of the fossil-fuel power station.

Determine the minimum annual mass loss of pure uranium-235 in the nuclear power station. [3]

d) A point on the rotor in the dynamo moves in a circle of radius 2.8 m. The rotor rotates 60 times a second.
Calculate the linear speed of the point on the rotor. [2]

*The following questions are for HL candidates **only**.*

6. Yellow light from a spectrum is incident normally on a diffraction grating of spacing 1.25 μm. The light has two close spectral lines of wavelengths 589 nm and 590 nm.

   a) Calculate the angular separation of the two spectral lines in the second-order spectrum. [3]
   b) Determine the number of slits of the diffraction grating that must be illuminated for the images of the two lines to be just resolved in the first-order spectrum. [3]
   c) The diffraction grating is illuminated by light from a single slit placed a few centimetres from the diffraction grating.
   Explain why there is a maximum width to this slit. [3]

7. a) State what is meant by *electric potential at a point*. [2]

   b) A point charge is travelling directly towards the centre of a charged metal sphere. The magnitude of the point charge is +42 nC and the sphere has a charge +3.5 μC.

   At one instant, the point charge is 2.5 m from the centre of the sphere and has a speed towards it of 1.8 m s^{-1}.

   i) Show that the electric potential of the sphere at a distance of 2.5 m from its centre is about 13 kV. [2]

   The electric potential of the sphere is 70 kV. The mass of the point charge is 0.23 g.

   ii) Determine whether the point charge will collide with the sphere. [4]

8. A metal aircraft is flying horizontally over the magnetic north pole at a constant velocity. At the pole, the magnetic field direction is vertically upwards.

   a) Explain how an emf is induced between the wing tips. [3]
   b) Faraday's law suggests that the induced emf is related to the rate of change of flux.
   Outline, for the aircraft, what is meant by *rate of change of flux*. [1]
   c) The speed of the aircraft relative to the ground is 270 m s^{-1} and its wingspan is 35 m. The vertical component of the magnetic field strength at the magnetic pole is 0.60 mT.
   Determine the induced emf across the wingspan. [2]
   d) Outline how Lenz's law applies to this situation. [3]

9. No photoelectron emission is observed from a particular metal surface when incident monochromatic light on it is below a minimum frequency.

   a) Outline why the wave theory for light cannot explain this observation. [2]

   b) Radiation of frequency 7.14×10^{14} Hz is incident on the metal surface. The work function of the metal is 2.0 eV.

      i) Calculate the threshold frequency of the metal. [2]

      ii) Determine the maximum kinetic energy, in eV, of the emitted electrons [4]

   c) The explanation by Einstein of the photoelectric effect is often described as a paradigm shift in physics.
      Outline what is meant by a *paradigm shift*. [1]

Paper 3

SL: 1 hour

HL: 1 hour 15 minutes

Instructions to candidates

- Answers must be written within the answer boxes provided (answer sheets are provided at **www.oxfordsecondary.com/ib-prepared-support**).
- A calculator is required for this paper.
- The maximum mark for the SL examination paper is **[35 marks]**.
- The maximum mark for the HL examination paper is **[45 marks]**.

Section A

*SL and HL candidates: answer **all** questions.*

1. A student obtains data showing the variation of pressure p with volume V for a gas. The data are given as a table of results and in the form of a graph. Error bars are shown on the graph.

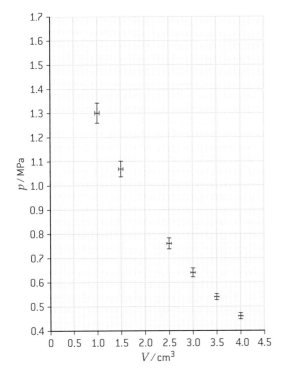

V / cm³	p / MPa
10	1.30
15	1.08
25	0.76
30	0.64
35	0.54
40	0.46

a) Estimate, using the graph, the percentage uncertainty in the value of p when $V = 1.0\,\text{cm}^3$. [2]

b) The student proposes that the data obey the relationship $p = \dfrac{k}{V}$, where k is a constant. Suggest, using the table, whether this relationship is correct. [3]

c) Estimate p when:
   i) $V = 2.0\,\text{cm}^3$ [1]
   ii) $V = 0.50\,\text{cm}^3$ [1]

d) Suggest which of your two estimates in part c) is more reliable. [2]

2. A student uses a metre ruler to measure the dimensions x and y of the horizontal surface of a small table. The readings the student took for x are shown in this table.

Reading	1	2	3	4	5
x / cm	49.5	50.7	49.8	50.2	49.9

    a) Calculate the mean value of x and its absolute uncertainty. [2]

    b) The mean value of y with its absolute uncertainty is (75.1 ± 0.4) cm.
    Determine the area of the table surface and its absolute uncertainty. [3]

    c) State **one** possible systematic error that can occur when using a metre ruler. [1]

Section B

*Answer **all** of the questions from **one** of the options.*

Option A: Relativity

*SL candidates: answer questions 3–6 **only**.*

HL candidates: answer questions 3–8.

3. A free electron moves parallel to a metal wire. Conduction electrons in the wire move with a drift speed equal to the speed of the free electron. An observer is stationary in the frame of reference of the wire.

    a) Outline what is meant by a *frame of reference*. [1]

    b) The observer describes the force acting on the electron as being magnetic.
    State and explain the nature of the force in the frame of reference of the free electron. [4]

4. A rocket approaches a space station at a speed relative to the station of $0.85c$. The proper length of the rocket is 650 m and the proper length of the station is 6.5 km.

    a) Calculate the length of the rocket according to an observer at rest in the space station. [2]

    b) A shuttle moves from the rocket to the station with a speed of $0.20c$ according to an observer at rest in the station. Calculate the velocity of the shuttle relative to the rocket. [2]

5. **a)** A pion decays in a proper time of 36 ns. It is moving with a velocity of $0.98c$ relative to an observer. Calculate the decay time of the pion as measured by the observer. [2]

    b) Explain, in terms of length contraction, the effect in part a). [3]

6. Jack and Jill are twins. Jack leaves Earth, according to Jill, at $0.6c$ to reach a position 3.0 light years away.

    a) Calculate the journey time according to:
      i) Jill
      ii) Jack. [3]

    b) Sketch a spacetime diagram in Jill's reference frame to show the worldlines for both Jack and Jill. [3]

7. An electron and a positron approach each other along the same line with the same speed. The initial kinetic energy of the electron is 2.0 MeV. In the collision, the electron and the positron are annihilated and two photons with identical energy are produced.

   a) Calculate the minimum speed of the electron. [3]

   b) Determine, for one photon, its:

   i) energy

   ii) momentum. [2]

8. **a)** A ball is thrown, initially horizontally, near to the surface of a planet.
   Explain, with reference to spacetime, the trajectory of the ball. [3]

   b) Calculate the radius that an object with the mass of the Earth would need in order to become a black hole. [2]
   Mass of Earth $= 6.0 \times 10^{24}$ kg.

Option B: Engineering physics

*SL candidates: answer questions 9–11 **only**.*

HL candidates: answer questions 9–13.

9. A revolving door rotates about a vertical axis completing one revolution in 18 s. When the motor is switched off, the system makes a further three complete revolutions while decelerating uniformly to rest. The total moment of inertia of the door is 8.0×10^3 kg m^2.

   a) Deduce the time taken for the system to come to rest. [3]

   b) Estimate the average frictional torque acting on the system. [3]

10. A solid cylinder and a hollow cylinder of the same mass and external radius are released from rest simultaneously at the top of a ramp. They roll down the ramp without sliding.

    Predict which cylinder will reach the bottom of the ramp first. [3]

11. A fixed mass of an ideal gas is taken through the following cycle of processes.

    ① Isothermal compression at low temperature.
    ② Constant volume increase in pressure.
    ③ Isothermal expansion at high temperature.
    ④ Constant volume cooling to the original pressure, volume and temperature.

    a) Sketch the complete cycle on a pV diagram. [3]

    b) Air in a cylinder of volume 7.0×10^{-5} m^3 has a pressure of 4.0×10^6 Pa and a temperature of 340 K.
    A piston at one end of the cylinder is now allowed to expand to a final volume of 1.3×10^{-4} m^3. At this point, the pressure in the cylinder is 1.4×10^6 Pa.

    i) Deduce whether or not the expansion of the gas is an adiabatic process. [3]

    ii) Show that the quantity of gas in the cylinder is about 0.1 mol. [2]

    iii) Calculate the final temperature of the gas. [2]

    iv) Outline the changes in the gas with reference to the first law of thermodynamics. [1]

The following questions are for HL candidates only.

12. A solid cube of wood with sides of 0.25 m has a density of 160 kg m^{-3}. It is placed in seawater of density 1030 kg m^{-3} and floats with one face horizontal.

    a) Estimate the fraction of the height of the cube that is initially under the surface. [2]

    b) The wood absorbs water with time. Assume that the dimensions of the cube do not change.
Predict the effect of this absorption on your answer to part a). [2]

13. A simple pendulum consists of a solid sphere suspended by a string from a support. The Q factor for the system is 150 and the angular frequency is 8.0 rad s^{-1}.

    a) Discuss the motion of the pendulum after it is set oscillating. [3]

    b) The pendulum support now oscillates horizontally with frequency f.
Compare the amplitude of the pendulum and its phase relative to the movement of the support when $f = 1.2$ Hz and $f = 3.0$ Hz. [3]

Option C: Imaging

SL candidates: answer questions 14–16 only.

HL candidates: answer questions 14–18.

14. Rays of light are incident on a diverging lens parallel to the principal axis. The wavefront corresponding to these rays has reached the surface of the lens.

    a) Describe, with a diagram, the passage of the wavefront through the lens. [2]

    b) Explain, in terms of the wavefront, how the lens subsequently forms a virtual image. [2]

    c) The diverging lens has a focal length of –6.0 cm. The wavefront passes from the diverging lens to a converging lens of focal length +18 cm. After passing through this second lens, the rays emerge parallel to the principal axis again.
Determine the distance between the lenses. [3]

15. A compound microscope is in normal adjustment.

    a) State the nature of the final image formed by the instrument. [1]

    b) The focal lengths of the objective and eyepiece lenses of the microscope are 30 mm and 90 mm respectively. The length of the microscope tube is 240 mm. An observer has a near point of 25 cm.
Determine the position of the object so that the image is at the observer's near point. [4]

16. An optical fibre has a core with refractive index $n = 1.56$ and cladding with $n = 1.38$.

    a) Calculate the critical angle for this core–cladding combination.   [2]

    b) A ray is incident on the end of the fibre with an angle of θ to the central axis of the core.
Calculate the maximum value of θ for which total internal reflection at the core–cladding interface occurs.   [3]

    c) A signal with a power of 15.0 mW is input to the optical fibre, which has a length of 5.70 km. The attenuation of this fibre is 1.24 dB km^{-1}.
Calculate the output power of the signal.   [3]

*The following questions are for HL candidates **only**.*

17. When X-rays travel through air of pressure p, in pascal, and temperature T, in kelvin, the half-value thickness is given in metres by

$$x_{\frac{1}{2}} = \frac{1.8 \times 10^5 \times T}{p}$$

X-rays reach the top of the atmosphere 25 km above the Earth's surface from space.
The average pressure of the atmosphere is 20 kPa and the average temperature is 240 K.

    a) Estimate the half-value thickness for the atmosphere.   [1]

    b) Determine the fraction of the X-ray intensity incident on the top of the atmosphere that is transmitted to the Earth's surface.   [2]

    c) Comment on the extent to which the atmosphere protects us on Earth from the incident X-rays at the top of the atmosphere.   [1]

18. a) Outline how protons are used in nuclear magnetic resonance.   [3]

    b) Explain the role of the gradient field in magnetic resonance imaging.   [3]

Option D: Astrophysics

*SL candidates: answer questions 19–22 **only**.*

HL candidates: answer questions 19–24.

19. The observed angular displacement of a star viewed from Earth over a six-month period is θ and the orbital diameter of the Earth is d. The Earth–star distance D.

    a) Draw a diagram to show the relationship between θ, d and D.   [1]

    b) When D is measured in parsec, θ is measured in arc-seconds. State one different consistent set of units for D and θ.   [1]

20. The star Zeta Puppis has a surface temperature of 42 400 K and a radius of 7.70×10^9 m. The parallax angle that Zeta Puppis subtends from Earth is 3.40×10^{-3} arc-seconds.

    a) Calculate the distance of Zeta Puppis from Earth.   [1]

    b) Deduce the luminosity of Zeta Puppis.   [2]

    c) Calculate the peak wavelength in the spectrum of Zeta Puppis.   [1]

21. A red giant star X has a luminosity 370 times that of the Sun.

    a) Determine the mass of X.
        Mass of Sun $= 2.0 \times 10^{30}$ kg. [2]
    b) Y is a red supergiant star.
        i)   Compare the likely evolution of Y to that of X. [2]
        ii) Suggest the circumstances in which the evolution of
             X and Y could be the same. [2]

22. The characteristics of three stars are given in this table.

	Luminosity	Surface temperature / K
Sun	3.8×10^{26} W $\equiv 1L_{\odot}$	5700
Capella	$80L_{\odot}$	5000
Vega	$40L_{\odot}$	9600

    a) Compare the surface area of Capella and the Sun. [3]
    b) Calculate the radius of Capella. [3]
    c) The apparent brightness of Vega is 2.2×10^{-8} W m^{-2}.
        Calculate, in parsec, the distance from Earth to Vega. [2]

*The following questions are for HL candidates **only**.*

23. a) Outline what is meant by the *Oppenheimer–Volkoff limit*. [1]
     b) Star A has a mass 100 times that of the Sun.
        i)   Predict, using your answer to part a), the evolution of A
             until it becomes a neutron star. [2]
        ii) Deduce the eventual outcome of A. [1]

24. a) Draw a graph showing the variation with time *t* of the cosmic
        scale factor *z* for:
        i)   a closed Universe without dark energy
        ii) an accelerating Universe with dark energy. [3]
     b) Explain how observations of supernovae have led to the
        suggestion that the Universe possesses dark energy. [3]